STATISTICAL PREDICTION ANALYSIS

STATISTICAL PREDICTION ANALYSIS

J. AITCHISON
Titular Professor in Statistics
University of Glasgow

I.R. DUNSMORE
Lecturer in Statistics
University of Sheffield

CAMBRIDGE UNIVERSITY PRESS
CAMBRIDGE
LONDON NEW YORK NEW ROCHELLE
MELBOURNE SYDNEY

CAMBRIDGE UNIVERSITY PRESS
Cambridge, New York, Melbourne, Madrid, Cape Town, Singapore, São Paulo, Delhi

Cambridge University Press
The Edinburgh Building, Cambridge CB2 8RU, UK

Published in the United States of America by Cambridge University Press, New York

www.cambridge.org
Information on this title: www.cambridge.org/9780521206921

First published 1975
First paperback edition 1980
Re-issued in this digitally printed version 2008

A catalogue record for this publication is available from the British Library

Library of Congress Catalogue Card Number: 74-25649

ISBN 978-0-521-20692-1 hardback
ISBN 978-0-521-29858-2 paperback

Contents

Preface

Prediction by its derivation (L. *praedicere*, to say before) means literally the stating beforehand of what will happen at some future time. It is an occupational hazard of many professions: meteorologist, doctor, economist, market researcher, engineering designer, politician and pollster. It is indeed a precarious game because any specific prediction can eventually be compared with the actuality. Many prophets of doom predicting that the world will end at 12.30 on 7 May are left in quieter mood by 12.31. Prediction is a problem simply because of the presence of uncertainty. Seldom, if ever, is it a case of logical deduction; almost inevitably it is a matter of induction or inference. Probabilistic and statistical tools are therefore necessary components of any scientific approach to the formalisation of prediction problems.

In this book we shall be concerned with prediction not only in this narrow sense of making a reasoned statement about what is likely to happen in some future situation but with a much wider class of problems. Any inferential problem whose solution depends on our envisaging some future occurrence will be termed a problem of statistical prediction analysis. The presentation in chapter 1 of a selection of motivating examples illustrates the nature and diversity of statistical prediction analysis, and serves as an introduction to the ingredients of the problem.

A science historian, writing on the development of the concepts and practice of prediction, would probably start by pointing out how primitive man was compelled to attempt prediction, for example the forecasting of the date on which the local river would flood. He might trace how traditions of prediction by magic and sorcery gave way to a realisation that past experience and observation often prove a reliable guide to future events, quoting as an example some well-known folk-lore rhyme about the weather, such as 'A red sky at night is a shepherd's delight'. He would move forward in time to the scientific revolution of the sixteenth and seventeenth centuries highlighting the great advances in descriptive astronomy culminating in what is possibly the greatest predictive achievement of all – the nautical almanack. He might then record the origins of the realisation that there is a relationship between the reliability of a prediction and the inherent variability of the data or the difficulty of accurate measurement, and he could indicate how the science of statistics in its sophisticated present-day form had emerged from such origins. Eventually he might feel the need for a chapter to explain the attitudes of present-day statisticians to the problem of prediction.

His enquiry into such attitudes would soon reveal an interesting though puzzling situation. If his first step were the natural one of taking stock of well-established theory and application of prediction by surveying statistical text-books he would be singularly disappointed. A search of their indexes would reveal only a small minority which listed the term 'prediction' or its near equivalents. Thorough reading of the texts would bring only a little further enlightenment. He would find a few which quoted a 'prediction interval' or a 'confidence interval for a future observation' towards the end of their discussion on regression analysis, where the regression data are to be used to indicate what is likely to happen when the basic experiment is performed at some specified value of the controllable variable. Moreover in very few texts would he find any clear statement of the principles on which such a prediction interval is based and of the method by which the interval is evolved from these principles. He might fare a little better with the small number of texts which present the 'tolerance interval' approach to prediction, but again he would be left with doubts about the basis, interpretation and usefulness of such intervals.

If he broadened his survey to include research literature and specialist books he would certainly find a fuller account of the tolerance interval approach but it is unlikely that all his doubts would be removed. He would also find a well-developed and still developing theory of prediction for stationary processes. On discovering this he might be forgiven for expressing surprise that a more fully developed theory apparently existed for this more complicated situation than for essentially simpler situations. Apart from these expositions he would find only a mixture of *ad hoc* techniques of forecasting by trend curves, exponential weighting, etc. He might then begin to wonder why it is that statisticians have devoted so much time, energy and skill to the fields of estimation, hypothesis-testing and experimental design to the comparative neglect of prediction analysis which is surely at the heart of many statistical applications.

Much of statistical analysis is concerned with making inferences about unknown distribution parameters. For example, given the outcomes from an experiment which is known to be $N(\mu, \sigma^2)$ is it reasonable to suppose that $\mu > 0$; what is a confidence interval for μ? Now the purpose of such inference statements is surely to convey to some second party information about what is likely to happen if the experiment is performed again, or perhaps repeated a number of times. It is surprising therefore that greater thought has not been given to the more direct practical type of inference, where statements are required for what is likely to occur when future experiments are performed. Indeed it is common practice for a statistician first to obtain from the experimental outcome an estimate of the indexing parameter of some class of distributions describing an experiment, and subsequently to use the estimate as if it were the true value to allow prediction. It is paradoxical that while the folly of this approach is pointed out in simple situations there is all too ready acceptance of it in more complicated situations.

This book is an attempt to present certain aspects of statistical prediction theory within a unified framework and notation. As we have already indicated statistical prediction analysis will be here considered in a wide sense, to include any form of statistical analysis where consideration of what may happen (or indeed, to be slightly esoteric at this early stage, what may already have happened) at performances of some future experiment or experiments is essential to the formulation of the problem. The development is considered from both a frequentist and a Bayesian viewpoint for it is our belief that there are situations which are essentially frequentist and other situations which are essentially Bayesian, and the particular type of analysis appropriate to the situation in hand should be used.

It is necessary to provide a comprehensive, yet clear, notation for all the possible distributions involved. Further the notation must not become over-elaborate, for example, when distinguishing between prior and posterior distributions. We believe we have achieved the necessary balance. The notation is introduced as required, mainly in chapters 1, 2 and 3, but a complete list is provided in appendix I.

We are grateful to Dr A.F. Lever of the Medical Research Council Blood Pressure Unit, Western Infirmary, Glasgow, for the data on Conn's syndrome and to Dr M. Damkjaer Nielsen of Glostrup Hospital, Copenhagen, for the data on Cushing's syndrome, used for illustrative purposes. The typing of the various drafts and the final version of this book was undertaken by Mrs. I.U. Adey, Miss E.M. Nisbet and Mrs M.S. Robertson. Their care, patience and good humour in the face of a continually changing manuscript played a major role in its eventual completion, and we wish to record our sincere thanks to them.

Glasgow, Sheffield J.A., I.R.D.

March 1975

1
Introduction

1.1 The nature of statistical prediction analysis

An essential feature of statistical prediction analysis is that it involves two experiments e and f. From the information which we gain from a performance of e, the *informative* experiment, we wish to make some reasoned statement concerning the performance of f, the *future* experiment. In order that e should provide information on f there must be some link between these two experiments. Throughout this book we shall deal with problems where this link is through the indexing parameter of the two experiments e and f, and so we make the following assumption.

Assumption 1 The class of probability models which form the possible descriptions of e and the class of possible models for f have the same index set Θ, and the true models have the same (though unknown) index θ^*.

A further general feature of all the problems we shall consider is contained in the following independence assumption.

Assumption 2 For given index θ the experiments e and f are independent.

By adopting this second assumption we deliberately exclude a range of prediction problems in which f is a continuation of some stochastic process of which e records a realisation to date. Techniques such as forecasting by exponential weighting, linear least squares prediction and time series analysis are thus outside the scope of this book.

To give some idea of the wide applicability of statistical prediction analysis as defined above and to motivate the development of appropriate theory we devote the remainder of this chapter to the presentation of specific prediction problems. All these problems are later analysed and extended in the sections indicated in the text.

1.2 Some examples

In its most direct form statistical prediction analysis may simply be the provision of some probabilistic statement about the likely outcome of the performance of f.

Table 1.1 *Survival times (weeks) of 20 carcinoma patients*

25	45	238	194	16	23	30	16	22	123
51	412	45	162	14	72	5	35	30	91

Example 1.1

Medical Prognosis. The data of table 1.1 are the survival times (weeks) of 20 patients presenting with a certain type of carcinoma and receiving treatment of preoperative radiotherapy followed by radical surgery. On the basis of this information what can appropriately be said about the future of a new patient with this type of carcinoma and assigned to this form of treatment? Clearly any rational statement would regard 100 weeks survival as much more plausible than 500 weeks survival, but how should such views be summarised and quantified? What is a reasonable assessment of the probability that the patient will survive 100 weeks?

In this example the informative experiment e consists of recording the survival times of the 20 patients already treated. The future experiment f consists of treating the new patient similarly and recording his survival time. If no change in the treatment has been made since the conducting of e, then e and f consist respectively of 20 replicates and a single replicate of the same basic trial (record the survival time of a treated patient) and are independent. Assumptions 1 and 2 are therefore satisfied.

Attempts to quantify medical prognoses are of vital importance when similar information on an alternative treatment, for example radical surgery followed by postoperative radiotherapy, is available and a choice has to be made between treatments for a particular patient.

A detailed analysis and developments of this example are given in §2.6.

There are many less direct forms of statistical analysis than that of providing probabilistic statements. For example the problem may be one of choosing between alternative courses of action. If the consequences of taking a course of action depend on the outcome of f, then we still technically describe the problem as one of statistical prediction analysis.

Example 1.2

Machine tool replacement. Table 1.2 shows the recorded lifetimes of 24 machine tools of a certain type. In a factory using one of these machine tools the question of the best inspection and replacement policy is under discussion. If the tool wears out while unattended there is a loss of $\xi = 1.8$ per minute until such time as it is inspected (and immediately replaced). To have an inspector in attendance at the machine tool costs $\eta = 2.4$ per minute. If the

Table 1.2 *Lifetimes (minutes) of 24 machine tools*

47	62	111	47	57	14	290	118	19	4	17	46
5	239	9	140	89	94	217	35	103	28	37	111

tool is replaced before it wears out an overhead cost $\zeta = 54$ is incurred and also a loss at rate ξ per minute of its unused lifetime has to be debited. What is the optimum policy on the basis of all the information?

Here again the experiments *e* and *f* are easily identified. If we regard a basic trial as consisting of the recording of the lifetime of a single machine tool then *e* and *f* consist of 24 replicates and a single replicate of the basic trial, and satisfy assumptions 1 and 2. Again we clearly wish to infer something about the performance of *f* (the current machine tool) from the information contained in *e*. But a statement about the relative plausibilities of the possible outcomes of *f* is not sufficient. We must decide on one of many possible courses of action, namely the time periods during which we wish an inspector to be present to investigate whether the tool is still functioning or has already broken down. Suppose for simplicity that the only courses of action open to us are to select a time, *a* say, at which to send in an inspector with instructions to replace the tool immediately. The consequences of taking action *a* depend on the outcome of *f*, the actual lifetime *y* of the current machine tool. For if *a* exceeds *y* there is lost production time $a - y$ with a corresponding loss of $\xi(a-y)$, whereas if *y* exceeds *a*, the overhead scrapping loss ζ is incurred together with a debit of $\xi(y-a)$ for unused productive capacity. Thus a prediction associated with the performance of the future experiment *f* is necessary for any rational analysis of the problem but the prediction is a means to the end of selecting an appropriate course of action. We have thus here a less direct form of statistical prediction analysis than in our previous example. The analysis of this problem is developed in §§3.2, 3.4.

Example 1.3

A quality control problem. Items are produced independently in large batches by a firm. The items may be either effective or defective and it is recognised by both manufacturer and customer that batches vary considerably in the number of effectives they contain. The terms of a suggested contract between manufacturer and customer require the manufacturer to test destructively 5 of the components of each batch. The remainder of the batch is to be supplied in packets of 25 with an accompanying statement about the maximum number of defectives each packet contains. The contract further requires that for at least 90 per cent of such batches the statement will be true for at least 80 per cent of packets. What statement strategy will fulfil the terms of this contract?

Here the future experiment *f* envisaged is the observation of the number of defectives in a packet of 25 components from a batch. The information that we have available consists of the observation of the number x of defectives in a sample of 5 components.

How we use the information in e to meet the requirements on f and how we interpret the 90 per cent and 80 per cent in the statements are discussed later in §6.3.

In the examples so far discussed we have attempted to illustrate some of the basic structure of statistical prediction analysis. The main feature emerging is the need to make some prediction of the outcome y of a future experiment f based on the outcome x of an informative experiment e. The relevance of the information obtained from e to the future experiment f is contained in what can conveniently be called the predictive density function. This concept is dealt with in detail in chapter 2 and is central to much of the subsequent analysis. For example the predictive density function provides the quantification of medical prognosis that we seek for example 1.1. In chapter 3 the introduction of utility functions to quantify the measures of gains or losses involved in making a prediction enables decisive prediction problems such as example 1.2 to be analysed in detail. Informative prediction problems in which no such measures are available, as in example 1.3, are dealt with in chapters 4, 5 and 6. A substantial part of these chapters presents the theory of tolerance regions from a fresh viewpoint. Some interesting relationships between decisive and informative prediction are developed in chapter 7, which also reviews other approaches such as empirical Bayes and distribution-free prediction. The remainder of the book is then devoted to specific areas of application and particularly to even more indirect forms of statistical prediction analysis. Some of these forms are now illustrated by examples.

1.3 Examples of choice of future experiment

There are many problems in which there is a whole class F of possible future experiments and the problem is to determine which future experiment f satisfies certain desirable properties. In choosing this experiment we have to envisage its performance and for this reason such problems fall within the scope of statistical prediction analysis. We consider here two representative examples.

Example 1.4

A problem of laminate design. In the manufacture of a laminate several sheets of material are superimposed. The sheets are liable to contain flaws and the total number y of flaws in the finished product can be measured by an X-ray device. The durability of the product increases as the number t of component sheets increases, but there is an upper limit y_0 to the total number of flaws which can be allowed before the product is rejected.

Table 1.3 *Numbers of flaws in nine laminate specimens*

Number of sheets	5	6	7	8	9	10	11	12	13
Number of flaws	3	2	4	5	7	6	7	7	8

In a pilot experiment 9 specimens of the product were made with 5, 6, 7, ... , 13 sheets of material superimposed with resulting flaws as shown in table 1.3. The management wishes to market as durable a product as possible but has decided that at most $y_0 = 7$ flaws can be allowed in any product. If the profit per component sheet for accepted products is ten times the cost per component sheet for rejected products how many sheets should be superimposed?

In order to resolve this problem we have to consider the number of flaws which may result if we use t sheets. Thus we are forced to envisage a whole class of possible experiments

$$F = \{f_t : t = 1, 2, \ldots\},$$

where f_t denotes the experiment of counting the total number of flaws in a laminate of t sheets. The problem is then to choose which future experiment f_t gives as high durability as possible and yet attempts to meet the flaw limitation. The information available comes from an informative experiment e yielding the data of table 1.3 and which could formally be written in the form

$$e = \{f_5, f_6, \ldots, f_{13}\},$$

a set of independent performances of $f_5, f_6, \ldots, f_{13}$. The direct prediction problem for f_t enters into our attempts to balance our desire to increase t (and hence the durability) and our concern that the number of flaws may increase beyond the acceptable limit. This *regulation* example is analysed in §9.3.

Example 1.5

Maximising yield of an industrial process. The yields (kg) shown in table 1.4 were obtained in an experiment in which an industrial process was run successively at 5 different temperatures and 3 different pressures, each combination of temperature and pressure being used twice. What combination of temperature and pressure should be used in order to maximise the yield in a future run of the process?

Here the problem is to determine at what combination $t = (t_1, t_2)$ of temperature t_1 and pressure t_2 to run the process. Denote by f_t or f_{t_1, t_2} the future experiment which records the yield from an operation of the industrial process at temperature t_1 and pressure t_2. We are thus forced to consider the class of future experiments

Table 1.4 *Yields (kg) from 30 process runs at different temperature-pressure combinations*

Temperature (°C)	Pressure (atmospheres) 1.00	1.25	1.50
50	65, 68	70, 72	73, 74
60	72, 70	75, 75	77, 76
70	73, 75	81, 83	79, 78
80	76, 75	81, 79	75, 77
90	76, 76	78, 80	76, 73

$$F = \{f_t : t \in T\},$$

where T denotes the set of possible temperature–pressure combinations. The informative experiment consists of 30 independent experiments (process runs) each of f_t type. To choose an experiment from the set F we must again envisage the outcomes of such future experiments and so are involved in statistical prediction analysis. This *optimisation* problem is analysed in §9.5.

1.4 Examples of detection of future experiment

A common statistical problem is to detect which one of a class F of 'future' experiments has already been performed from the information from e and the known outcome y of the performed future experiment. While in such circumstances it may seem strange to use the term statistical prediction analysis we shall see that we are led inevitably to the same concepts of prediction as we have already encountered. Indeed we have to envisage prediction for each of the possible future experiments in F. Two examples illustrate the nature of the problem here.

Example 1.6

Antibiotic assay. When a droplet of specified volume of an antibiotic is placed on an infected medium on a Petri dish and kept under controlled conditions the antibiotic clears a circular area of the medium. Moreover the diameter of the cleared area depends on the concentration of the antibiotic although this relationship is not a deterministic one. The idea underlying a biological assay of the *unknown* concentration of antibiotic in a blood specimen from a patient is to place droplets of standard antibiotic at different *known* concentrations and droplets from the specimen on the same batch of infected medium (fig. 1.1). The problem is then to infer from the relative sizes of the diameters associated with droplets of known concentration and of the diameters associated with droplets of the unknown concentration as much as possible about the unknown concentration. Such a direct comparison between the patient's specimen and the standard is usually necessary because the relationship between diameter and concentration usually varies from batch to batch of the medium.

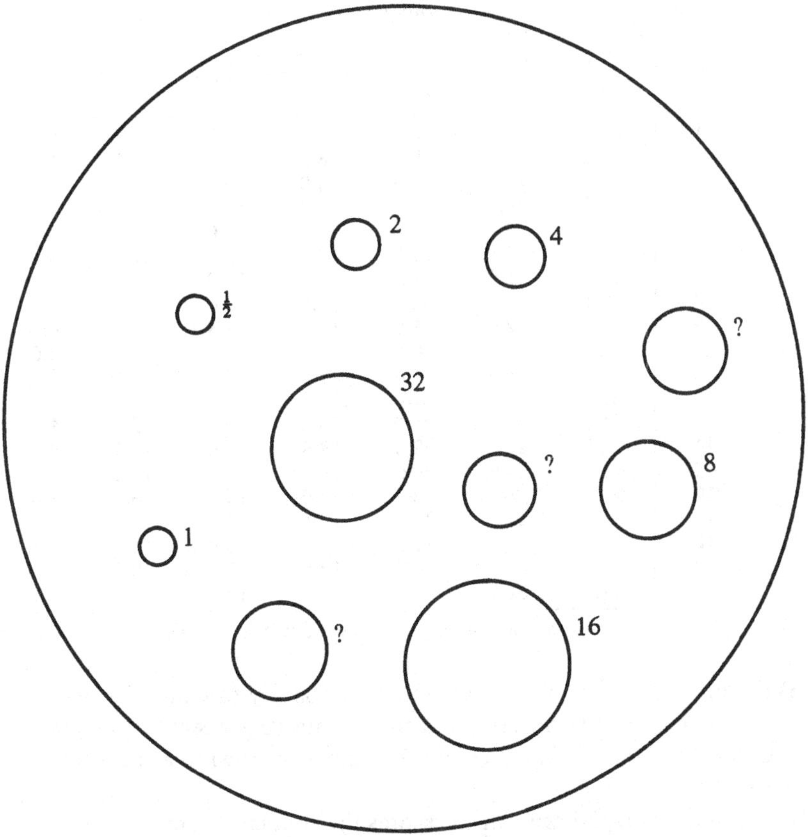

Fig. 1.1 Typical clearance circles in an antibiotic assay. For circles from standard droplets the known concentrations (mcg/ml) are shown. The three circles labelled ? are from droplets from a single specimen of unknown concentration.

Table 1.5 shows the results of an experiment to investigate the feasibility and reliability of this type of assay for a particular antibiotic. From the same batch of infected medium 20 Petri dishes were prepared and on each one droplet at each of seven concentrations (mcg/ml) was placed and the resulting clearance diameters recorded. Typical questions that we have to be in a position to answer are the following.

(1) Suppose that a droplet from a particular specimen has been placed on medium from this batch and has cleared a circle of diameter 19 mm. What can we infer about the concentration of antibiotic in the specimen?

(2) If three droplets from the same blood specimen have given clearance diameters of 18.0, 19.5 and 19.5 mm, what can be inferred about the concentration of the antibiotic in the specimen?

Table 1.5 *Clearance diameters (mm) on 20 Petri dishes for different concentrations (mcg/ml) of antibiotic*

Dish no.	Concentration (mcg/ml) 32	16	8	4	2	1	0.5
1	21.5	22.5	22.0	20.0	15.0	10.0	*
2	24.5	19.0	19.0	18.5	17.0	10.0	*
3	21.0	22.0	20.0	18.0	16.5	12.0	5.0
4	20.5	20.5	21.0	19.0	17.0	7.0	*
5	21.5	23.0	20.0	19.5	18.0	11.0	7.0
6	22.0	21.0	21.0	19.0	16.0	12.0	5.0
7	21.0	21.0	21.5	19.5	17.0	8.0	*
8	20.5	22.0	20.5	19.5	17.0	11.0	9.0
9	21.5	22.0	21.5	18.0	17.5	13.0	6.5
10	22.5	22.5	20.5	18.5	16.0	11.5	4.0
11	22.5	21.0	20.0	19.0	17.0	9.5	*
12	22.5	21.5	22.0	21.5	18.0	11.0	*
13	21.5	22.0	20.0	20.0	19.5	12.0	*
14	22.0	21.0	22.0	19.5	16.0	12.5	*
15	23.5	23.0	20.0	19.5	17.0	11.5	*
16	21.5	21.0	22.0	19.0	16.0	6.5	5.0
17	21.0	22.0	20.5	19.5	18.0	11.0	*
18	22.5	22.0	21.0	19.0	16.0	11.5	*
19	22.0	22.0	20.5	18.5	17.0	10.0	*
20	21.0	22.5	21.5	20.5	17.5	11.0	7.0

An entry * indicates that no measurable clearance was achieved.

(3) For medical reasons we may wish to be reasonably sure that the concentration quoted for a particular specimen is within 10 per cent of the true value. How many droplets from the specimen should be used to achieve this reliability?

Let f_t denote an experiment which records the clearance diameter of a droplet of concentration t of the antibiotic, and consider the class

$$F = \{f_t : t \in T\}$$

of experiments indexed by the set T of all possible concentrations. Then the informative experiment e which yields the data of table 1.5 is clearly of regression type and may be expressed briefly as

$$e = \{f_{t_1}, \dots, f_{t_{140}}\},$$

a set of 140 independent experiments, where $t_1, \dots, t_{140}$ are the concentrations associated with the 140 droplets. The data from this informative experiment thus consist of 140 pairs $(t_1, x_1), \dots, (t_{140}, x_{140})$ of concentrations and corresponding clearance diameters, and can thus be set out in a typical scatter diagram (fig 1.2). For convenience a logarithmic concentration scale has been used.

In considering the first question posed we have the outcome $y = 19$ of some experiment from the class F, say f_u where u is the unknown concentration.

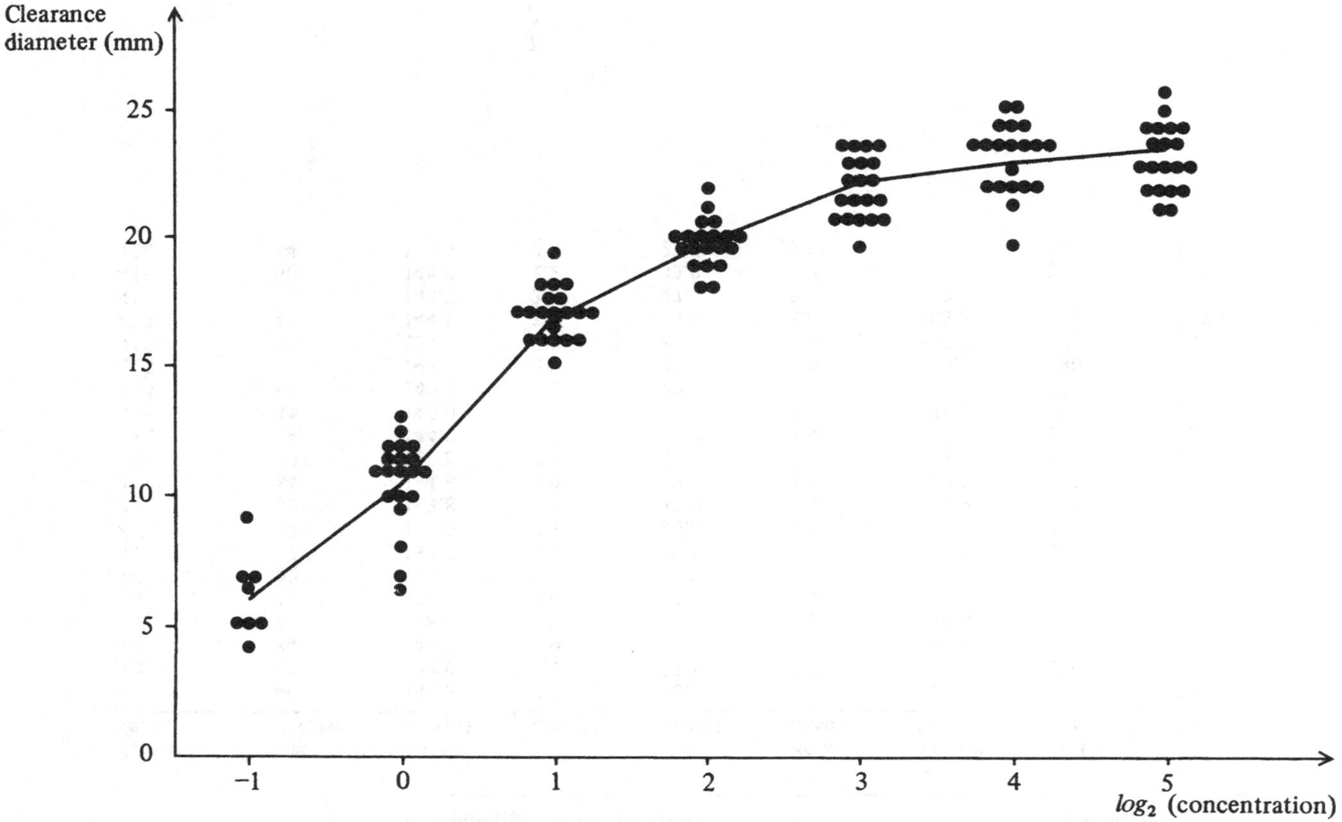

Fig. 1.2 Variability of clearance diameter with concentration of antibiotic. The line shown joins the mean clearance diameters at the different concentrations.

Table 1.6 *Results of eight preoperative tests on patients with Conn's syndrome*

		Concentrations in blood plasma					Blood Pressures	
Patient no.	Age (Years)	Na (meq/l)	K (meq/l)	CO_2 (meq/l)	Renin (meq/l)	Aldo-sterone (meq/l)	Systolic (mm Hg)	Diastolic (mm Hg)
*A*1	40	140.6	2.3	30.3	4.6	121.0	192	107
*A*2	37	143.0	3.1	27.1	4.5	15.0	230	150
*A*3	34	140.0	3.0	27.0	0.7	19.5	200	130
*A*4	48	146.0	2.8	33.0	3.3	30.0	213	125
*A*5	41	138.7	3.6	24.1	4.9	20.1	163	106
*A*6	22	143.7	3.1	28.0	4.2	33.0	190	130
*A*7	27	137.3	2.5	29.6	5.4	52.1	220	140
*A*8	18	141.0	2.5	30.0	2.5	50.2	210	135
*A*9	53	143.8	2.4	32.2	1.5	68.9	160	105
*A*10	54	144.6	2.9	29.5	3.0	144.7	213	135
*A*11	50	139.5	2.3	26.0	2.6	31.2	205	125
*A*12	44	144.0	2.2	33.7	3.9	65.1	263	133
*A*13	44	145.0	2.7	33.0	4.1	38.0	203	115
*A*14	66	140.2	3.1	29.1	4.7	43.1	195	115
*A*15	39	144.7	2.9	27.4	0.9	65.1	180	120
*A*16	46	139.0	3.1	31.4	2.8	192.7	228	133
*A*17	48	144.8	1.9	33.5	3.8	103.5	205	132
*A*18	38	145.7	3.7	27.4	2.8	42.6	203	117
*A*19	60	144.0	2.2	33.0	3.2	92.0	220	120
*A*20	44	143.5	2.7	27.5	3.6	74.5	210	114

(*continued*)

Table 1.6 (*continued*)

Patient no.	Age (Years)	Concentrations in blood plasma Na (meq/l)	K (meq/l)	CO_2 (meq/l)	Renin (meq/l)	Aldo-sterone (meq/l)	Blood Pressures Systolic (mm Hg)	Diastolic (mm Hg)
*B*1	46	140.3	4.3	23.4	6.4	27.0	270	160
*B*2	35	141.0	3.2	25.0	8.8	26.3	210	130
*B*3	50	141.2	3.6	25.8	4.1	20.9	181	113
*B*4	41	142.0	3.0	22.0	4.7	20.4	260	160
*B*5	57	143.5	4.2	27.8	4.3	23.7	185	125
*B*6	57	139.7	3.4	28.0	5.2	46.0	240	130
*B*7	48	141.1	3.6	25.0	2.5	37.3	197	120
*B*8	60	141.0	3.8	26.0	6.5	23.4	211	118
*B*9	52	140.5	3.3	27.0	4.2	24.0	168	104
*B*10	49	140.0	3.6	26.0	6.3	39.8	220	120
*B*11	49	140.0	4.4	25.6	5.1	47.0	190	125
Undiagnosed	50	143.3	3.2	27.0	8.5	51.0	210	130

A: adenoma
B: bilateral hyperplasia

The problem here differs from that of previous examples in that the 'future' experiment has already been performed; we know its outcome but we do not know which experiment in the class F has been performed. We have thus a kind of inverse prediction problem, but we shall find in its solution that the same concepts of statistical prediction analysis arise as in more direct problems.

The second and third questions are very similar in nature. To deal with the second question we have the outcome (y_1, y_2, y_3) of the 'future' experiment $\{f_u, f_u, f_u\}$ consisting of three replicates of f_u, and again we face the problem of detecting what f_u has been performed. To analyse the third question we have to envisage the consequences of the 'future' experiment consisting of K replicates of some f_u and investigate how these consequences vary with K.

All these questions, which fall within the general class of *calibration* problems, are fully analysed in § 10.8.

Example 1.7

Preoperative medical diagnosis. Recently at operation it has been discovered that a rare syndrome of hypertension (Conn's syndrome) can be due to either (1) a benign tumour (adenoma) in the adrenal cortex or (2) a more diffuse condition (bilateral hyperplasia) of the adrenal glands. The assessment of treatment, which may range from total adrenalectomy, through removal of an adenomatous adrenal gland if locatable, to drug therapy, is now recognised to depend on the diagnostic assessment and on a number of factors external to the diagnostic assessment. It is therefore highly desirable to be in a position to obtain preoperatively a reasonable diagnostic assessment of the relative plausibilities of the two types. Since radiology is not yet a reliable diagnostic tool in this differential diagnostic problem, the possibility of diagnosing preoperatively from eight diagnostic tests has been considered. Table 1.6 shows the results of these eight preoperative tests on 31 patients who have in earlier years been operated on and their type, 1 or 2, definitely established. The eight tests have now been carried out on a new patient known to have Conn's syndrome but of as yet undiagnosed form. The results of these eight tests are shown in the final row of table 1.6.

The reader may recognise this problem under some different statistical heading such as discriminant analysis. We shall show that we gain both in simplicity and in practical application by considering it within the unifying framework of statistical prediction analysis. Indeed its form is similar to that of example 1.6, the only difference being that in the calibration example the index set T for the class F is continuous whereas in the present context T is discrete. Let f_t denote the experiment of recording the eight test results of a patient in category t; and let

$$F = \{f_t : t = 1, 2\},$$

so that $T = \{1, 2\}$ is the index set of the class F. If we regard our 'future' experiment f as the recording of the eight test results for the new patient then while we know the outcome of the future experiment we do not know which experiment of F has been performed. As in the calibration problem we shall find that prediction plays a central role in the satisfactory analysis of this diagnostic problem. The informative experiment e which yields the data of table 1.6 consists essentially of 20 replicates (with different adenoma patients) of f_1 and 11 replicates (with different bilateral hyperplasia patients) of f_2.

For the further analysis of this *diagnostic* problem see §11.6.

Problems

For each of the following prediction problems identify the informative experiment e and the future experiment f. These problems will be set for fuller analysis in later chapters, and there is no need to attempt to resolve them or to answer the questions posed.

1.1 The biparietal diameter of the foetal head can be measured by ultrasonic cephalometry while the baby is still in the mother's womb. This measurement gives a good indication of the size of the baby. To allow for the possibility of inducing labour prematurely if the baby is not developing, it would be useful to have some indication of the range of likely values of the diameter during the 7th month of pregnancy for normal babies. Provide such a range of normality if the measurements shown below are those taken during the 7th month for 50 normal babies who were born at full term.

Biparietal diameter (cm)

8.91	9.53	9.23	8.04	8.42	8.85	7.51	8.59	8.06	8.93
8.98	9.25	8.80	8.61	8.62	8.84	8.24	8.16	9.20	8.10
9.02	8.15	8.06	8.71	8.91	9.08	8.29	8.71	8.88	7.96
9.46	8.93	8.54	9.78	8.39	8.65	7.94	8.48	8.76	9.20
8.78	8.86	9.00	9.21	8.44	9.20	8.58	8.98	8.76	9.06

1.2 Close study of ten patients discharged from hospital after treatment for a chronic disorder has been undertaken and the ten observed times to first relapse found to have average 100 days.

The clinic is now trying to formulate a reasonable policy for recall time of patients which, for administrative purposes, must be the same for all patients. On the assumption that times to relapse have an exponential distribution what recall time for future patients would you suggest to ensure that 95 per cent of patients are recalled for retreatment before relapse?

A further attempt to rationalise policy suggests that the disutility for any patient who relapses before recall is 3 units for each day until recall day, whereas for patients who have not relapsed by recall day the disutility can be regarded as 1 unit for each day by which recall precedes the day on which relapse would have taken place. How does this affect the previous policy? What is the expected disutility per patient with this policy?

1.3 An income tax department has been asked to investigate the consequences of changing from a tax based on personal income to one based on household income, and to set a minimum taxable household income so that at least 25 per cent of households are exempt from the tax. The incomes (in £s) of 50 randomly selected households are as shown below; and previous studies suggest that logarithms of household incomes are normally distributed. The department likes to be 90 per cent sure of any statement it issues. What minimum taxable household income should be quoted?

3350	3170	1230	2420	2960	3250	3150	990	1870	1120
3620	5340	1810	1980	3360	2490	3310	1730	2090	4270
1510	2850	1790	3080	3880	1802	2190	1440	1250	1740
3510	2680	3880	2340	1560	2150	2070	1520	4340	2230
2100	2080	4590	2960	3510	2620	5250	2540	3860	1720

1.4 In a study on the setting strength of a woodwork adhesive the following procedure is carried out. Adhesive is applied to two strips of wood. After an interval t_1 the two strips are clamped together at right angles for time t_2. Then the force x (in kgs) required to prize the strips apart is measured. In the study the two factors t_1 and t_2 are varied as shown, 48 experiments being performed following the prescribed procedure for 3 repetitions of each of the 16 possible pairings.

What values of t_1, t_2 should the manufacturer recommend if the user wishes to maximise the strength of his joints?

	t_2 (hrs)			
t_1 (mins)	12	24	36	48
5	26.1	31.3	35.8	35.8
	28.5	31.4	36.3	36.0
	29.4	32.0	39.5	37.8
10	30.9	35.3	40.7	40.0
	30.9	36.3	41.8	40.7
	31.2	40.4	43.0	42.9
15	35.1	42.0	46.0	44.3
	39.7	44.1	48.1	45.1
	40.3	47.0	49.9	46.2
20	36.1	42.0	41.1	37.3
	36.9	43.1	41.9	38.2
	37.7	43.3	44.0	40.1

1.5 The technique of radiocarbon dating can be used to date archaeological specimens. The amount of radiocarbon in the specimen is determined by measuring the rate of disintegration of Carbon 14 in the specimen. From this measurement a radiocarbon age can be evaluated. A calibration curve is obtained by assessing the radiocarbon ages of samples of the bristlecone pine tree. The true ages of these samples can be accurately determined from the tree ring markings, and the bristlecone pine tree is particularly suitable for the purpose since the species is capable of survival for over 4500 years. The data are shown below for the relevant period of history.

A newly found archaeological specimen has a radiocarbon age of 4010 years. What is its true age if prior archaeological knowledge suggests a value in the region 3700 to 4500 years old?

Radiocarbon age (yrs)	3300	3500	3690	3510	3800	3800
Bristlecone pine age (yrs)	3500	3600	3720	3750	3910	4030
Radiocarbon age (yrs)	4090	3910	4110	4000	4200	4210
Bristlecone pine age (yrs)	4210	4220	4370	4440	4600	4620

1.6 When running under control a complex chemical process yields an output for which two quantitative characteristics are positively correlated. There are two points *a* and *b* at which the process may run out of control, but it is difficult to determine which without costly dismantling. When the process runs out of control the correlation between the characteristics is thought to be less strong or even negative. In the most recent 26 production runs after each of which dismantling of *a* and *b* was carried out to determine any loss of control results were obtained as in the table below.

The problem is to decide on the basis of the characteristics observed for the output from a new run whether to investigate at points *a* and *b* before the next run. The table below shows the losses involved in the various (action, fault) combinations.

Loss of control	Observed characteristics
None	(36.5, 66.4), (29.8, 61.7), (33.0, 63.8), (39.0, 68.2), (35.0, 66.7) (35.8, 67.4), (40.7, 72.2), (35.5, 68.4), (34.8, 65.2), (37.0, 67.0) (40.7, 70.6), (33.4, 65.3), (34.5, 68.0), (31.3, 63.4), (38.9, 71.2)
At *a* only	(38.6, 66.5), (37.4, 67.3), (38.8, 65.4)
At *b* only	(33.5, 67.4), (34.2, 70.5), (32.6, 69.4), (34.1, 68.9), (35.5, 69.2)
At both *a* and *b*	(32.8, 68.2), (32.2, 66.8), (33.1, 66.5)

Action taken	Loss of control None	At *a* only	At *b* only	At both *a* and *b*
No dismantling	0	5	6	9
Dismantle *a*	2	0	4	3
Dismantle *b*	3	4	0	3
Dismantle both *a* and *b*	4	3	3	0

2
Predictive distributions

2.1 Notation and terminology

Throughout this book we make use of the increasingly familiar device of labelling functions by their arguments. Suppose that we specify a density function $p(\theta)$ on a space Θ and that, for each θ in Θ, we specify a density function $p(y|\theta)$ on a space Y. The vertical bar preceding θ denotes conditioning on this known value of the index. The absolute or unconditional density function $p(y)$ on Y is then given by

$$p(y) = \int_{\Theta} p(y|\theta)p(\theta)\mathrm{d}\theta. \tag{2.1}$$

Note that $p(\theta)$ and $p(y)$ are completely different functions. We shall never have to attach numerical values to θ and y in the theoretical development; these symbols will therefore always be present to ensure the distinction between the two functions. No confusion arises from this device and the advantages of it are overwhelming. By its use we can express concepts with greater clarity and can avoid overdecorating the structure of the analysis.

We use the term *density function* not only in its usual context of describing a spread of the unit of probability over a continuous space but also to describe the breaking up of the unit of probability into discrete pieces on a discrete space. For example we speak not only of the exponential density function

$$p(y|\theta) = \theta \exp(-\theta y) \quad (y > 0)$$

on the positive real line, but also of the Poisson density function

$$p(y|\theta) = \theta^y \exp(-\theta)/y! \quad (y = 0, 1, 2, \ldots)$$

on the set of non-negative integers. Note also the device of stating the effective sample space, the domain of non-zero probability density, within brackets in the specification of the density function. Moreover, in theoretical developments where spaces may be either continuous or discrete, we always use the integral sign $\int$, as in (2.1), to indicate the mathematical process of accumulation, whether the process is integration or summation.

For standard distributions a mnemonic notation, which will allow complicated results to be simply stated and readily assimilated, will be introduced as it is required. For reference purposes, however, appendix I provides a complete list of this notation.

2.2 Two sources of information and their combination

Suppose that there is some future experiment f in which we are interested, and that for its probabilistic description we have a sample space Y and a class of possible density functions

$$\{p(y|\theta):\theta \in \Theta\}$$

on Y. The indexing set Θ is assumed to be known but the true index θ^* is not known. This uncertainty about the true index is the source of the statistical nature of the prediction problem. If we knew θ^* we would know precisely the density function $p(y|\theta^*)$ describing our uncertainty about the outcome of the future experiment f. And we could do no better than this.

Our uncertainty about the true index may be alleviated by information from two sources.

(i) *Prior information about* θ. At this stage of statistical prediction analysis we adopt a Bayesian approach. We assume that while we do not know the true index θ^* we can place plausibilities or probabilities on the various possible θ in Θ. More precisely we assume that we have a known density function $p(\theta)$ on Θ. Later in chapters 4, 5 and 6 we shall examine the considerable alteration and complications to the statistical analysis when such information is not available.

(ii) *Informative experiment*. We suppose that it is possible to perform some informative experiment e with sample space X. The way in which the outcome of such an experiment conveys information about the true index is through assumption 1 of § 1.1, that the index set of the class of possible density functions describing e is also Θ and that the true index operating is θ^*, the same as for f. The class of density functions for e can then be denoted by

$$\{p(x|\theta):\theta \in \Theta\}$$

on X.

The information we obtain by observing x at the performance of the informative experiment e clearly influences the prior plausibilities $p(\theta)$ we have attached to the possible indices. The updating of the plausibilities in the light of the observation x is effected by the mechanism of Bayes's theorem, which in our notation leads to the posterior plausibility function

$$p(\theta|x) = \frac{p(\theta)p(x|\theta)}{p(x)}, \tag{2.2}$$

where

$$p(x) = \int_\Theta p(\theta)p(x|\theta)\,d\theta. \tag{2.3}$$

We introduce in table 2.1 some notation for the standard probability distributions which may apply to the informative experiment, together with the conjugate prior distributions for the parameters. We emphasise again that a *complete* list of the notation used is contained in appendix I, and therein can be found the meanings of such symbols as $D(g, h)$, $\mathcal{S}^d$, Γ_d, etc., which are used in table 2.1. The choice of family of prior plausibility functions for any particular situation is inevitably a nice judgement between mathematical tractability and practical considerations, and we adopt the now familiar Bayesian device of using the rich families of conjugate prior distributions. By choosing such conjugate prior distributions we ensure that the prior and the posterior distributions are both of the same family. Also shown in the final column of the table are the particular members of these families which correspond to vague prior information. The simple rules of updating from prior to posterior plausibility function are shown in table 2.3 for the standard distributions; the use of this table is explained in §2.4.

2.3 The predictive density function

So far we have used the information available to make a plausibility assessment about the unknown index. But this is not our final objective. We are interested in the outcome y of the future experiment f and so we want to use our assessment about the plausibility of indices to induce the plausibilities attaching to the unknown outcome. We are uncertain about which density function $p(y|\theta)$ applies to f but we have assessed the plausibility $p(\theta|x)$ of θ. The essence of the Bayesian approach to prediction problems is the concept of the probability distribution of a future observation y given the outcome x of the informative experiment. Clearly the plausibility of y given $p(\theta)$ and x is expressed by

$$p(y|x) = \int_\Theta p(y|\theta)p(\theta|x)d\theta, \tag{2.4}$$

and we term this function on Y the *predictive density function* $p(y|x)$ for y given $p(\theta)$ and x. The concept has been growing in popularity recently and is at the heart of much of the subsequent analysis of this book. Because of this importance we provide a formal definition.

Definition 2.1

Predictive density function. For a future experiment f with class of density functions

$$\{p(y|\theta):\theta \in \Theta\} \text{ on } Y,$$

Table 2.1 *Standard distributions and their conjugate priors*

Notation and name	Random variable and domain	Probability density function	Parameter restrictions [Vague prior information]
1 Binomial			
$\mathrm{Bi}(n,\theta)$ Binomial	x $x = 0, 1, \ldots, n$	$\binom{n}{x}\theta^{x}(1-\theta)^{n-x}$	$0 \leqslant \theta \leqslant 1$
$\mathrm{Be}(g,h)$ Beta	θ $0 \leqslant \theta \leqslant 1$	$\dfrac{\theta^{g-1}(1-\theta)^{h-1}}{\mathrm{B}(g,h)}$	$g > 0, h > 0$ $[g \to 0, h \to 0]$
2 Poisson			
$\mathrm{Po}(\theta)$ Poisson	x $x = 0, 1, 2, \ldots$	$\dfrac{\theta^{x}\exp(-\theta)}{x!}$	$\theta > 0$
$\mathrm{Ga}(g,h)$ Gamma	θ $\theta > 0$	$\dfrac{h^{g}\theta^{g-1}\exp(-h\theta)}{\Gamma(g)}$	$g > 0, h > 0$ $[g \to 0, h \to 0]$
3 Gamma			
$\mathrm{Ga}(k,\theta)$ Gamma	x $x > 0$	$\dfrac{\theta^{k}x^{k-1}\exp(-\theta x)}{\Gamma(k)}$	$\theta > 0, k > 0$
$\mathrm{Ga}(g,h)$ Gamma	θ $\theta > 0$	$\dfrac{h^{g}\theta^{g-1}\exp(-h\theta)}{\Gamma(g)}$	$g > 0, h > 0$ $[g \to 0, h \to 0]$
$\mathrm{Ex}(\theta) = \mathrm{Ga}(1,\theta)$ Exponential	x $x > 0$	$\theta\exp(-\theta x)$	$\theta > 0$

Table 2.1 (*continued*)

Notation and name	Random variable and domain	Probability density function	Parameter restrictions [Vague prior information]
4 Multinomial			
$\mathrm{Mu}(n, \boldsymbol{\theta})$ Multinomial	$\mathbf{x} = (x_1, \ldots, x_d)$ $x_i = 0, 1, \ldots, n,$ $\Sigma x_i \leqslant n$	$\binom{n}{\mathbf{x}} \theta_1^{x_1} \ldots \theta_d^{x_d} (1 - \Sigma \theta_i)^{n - \Sigma x_i}$	$0 \leqslant \theta_i \leqslant 1$ $\Sigma \theta_i \leqslant 1$
$\mathrm{Di}(\mathbf{g}, h)$ Dirichlet	$\boldsymbol{\theta} = (\theta_1, \ldots, \theta_d)$ $0 \leqslant \theta_i \leqslant 1,$ $\Sigma \theta_i \leqslant 1$	$\frac{1}{D(\mathbf{g}, h)} \theta_1^{g_1 - 1} \ldots \theta_d^{g_d - 1} (1 - \Sigma \theta_i)^{h-1}$	$g_i > 0, h > 0$ $[g_i \to 0, h \to 0]$
5 Normal			
$\mathrm{No}(\mu, \tau)$ Normal	x $x \in \mathbf{R}^1$	$\left(\frac{\tau}{2\pi}\right)^{1/2} \exp\{-\frac{1}{2} \tau (x - \mu)^2\}$	$\mu \in \mathbf{R}^1, \tau > 0$
$\mathrm{NoCh}(b, c, g, h)$ Normal chi-squared	(μ, τ) $\mu \in \mathbf{R}^1, \tau > 0$	$p(\mu \mid \tau)$ is $\mathrm{No}(b, c\tau)$ $p(\tau)$ is $\mathrm{Ch}(g, h)$: $\frac{(\frac{1}{2}h)^{g/2} \tau^{(g/2)-1} \exp(-\frac{1}{2}h\tau)}{\Gamma(\frac{1}{2}g)}$	$b \in \mathbf{R}^1, c > 0,$ $g > 0, h > 0$ $[c \to 0, g \to 0, h \to 0]$
6 Multinormal			
$\mathrm{No}_d(\boldsymbol{\mu}, \boldsymbol{\tau})$ Normal	$\mathbf{x} = (x_1, \ldots, x_d)$ $\mathbf{x} \in \mathbf{R}^d$	$\frac{\lvert \boldsymbol{\tau} \rvert^{1/2}}{(2\pi)^{d/2}} \exp\{-\frac{1}{2}(\mathbf{x} - \boldsymbol{\mu})' \boldsymbol{\tau} (\mathbf{x} - \boldsymbol{\mu})\}$	$\boldsymbol{\mu} \in \mathbf{R}^d$ $\boldsymbol{\tau} \in \mathcal{S}^d$
$\mathrm{NoWi}_d(\mathbf{b}, c, g, \mathbf{h})$ Normal-Wishart	$(\boldsymbol{\mu}, \boldsymbol{\tau})$ $\boldsymbol{\mu} \in \mathbf{R}^d, \boldsymbol{\tau} \in \mathcal{S}^d$	$p(\boldsymbol{\mu} \mid \boldsymbol{\tau})$ is $\mathrm{No}_d(\mathbf{b}, c\boldsymbol{\tau})$ $p(\boldsymbol{\tau})$ is $\mathrm{Wi}_d(g, \mathbf{h})$: $\frac{\lvert \frac{1}{2}\mathbf{h} \rvert^{g/2} \lvert \boldsymbol{\tau} \rvert^{(g-d-1)/2} \exp(-\frac{1}{2} \mathrm{tr}\, \mathbf{h} \boldsymbol{\tau})}{\Gamma_d(\frac{1}{2}g)}$	$\mathbf{b} \in \mathbf{R}^d, c > 0,$ $g > d - 1, \mathbf{h} \in \mathcal{S}^d$ $[c \to 0, g \to 0, \mathbf{h} \to 0]$

Table 2.1 (*continued*)

Notation and name	Random variable and domain	Probability density function	Parameter restrictions [Vague prior information]
7 Exponential (two-parameter)			
El(μ, τ) Exponential left-sided	x $x < \mu$	$\tau \exp\{-\tau(\mu - x)\}$	$\mu \in R^1, \tau > 0$
ErGa(b, c, g, h) Exponential (right-sided)-gamma	(μ, τ) $\mu \in R^1, \tau > 0$	$p(\mu\|\tau)$ is Er($b, c\tau$) $p(\tau)$ is Ga(g, h)	$b \in R^1, c > 0,$ $g > 0, h > 0$ $[b \to -\infty, c \to 0, g \to 0, h \to 0]$
Er(μ, τ) Exponential right-sided	x $x \geqslant \mu$	$\tau \exp\{-\tau(x - \mu)\}$	$\mu \in R^1, \tau > 0$
ElGa(b, c, g, h) Exponential (left-sided)-gamma	(μ, τ) $\mu \in R^1, \tau > 0$	$p(\mu\|\tau)$ is El($b, c\tau$) $p(\tau)$ is Ga(g, h)	$b \in R^1, c > 0,$ $g > 0, h > 0$ $[b \to \infty, c \to 0, g \to 0, h \to 0]$

an informative experiment e with class of density functions

$$\{p(x\,|\,\theta) : \theta \in \Theta\} \text{ on } X,$$

and a prior density function $p(\theta)$ on Θ, the *predictive density function* $p(y\,|\,x)$ for f is defined by

$$p(y\,|\,x) = \int_\Theta p(y\,|\,\theta)\,p(\theta\,|\,x)\,\mathrm{d}\theta$$

$$= \frac{\int_\Theta p(y\,|\,\theta)\,p(x\,|\,\theta)\,p(\theta)\,\mathrm{d}\theta}{\int_\Theta p(x\,|\,\theta)\,p(\theta)\,\mathrm{d}\theta}. \tag{2.5}$$

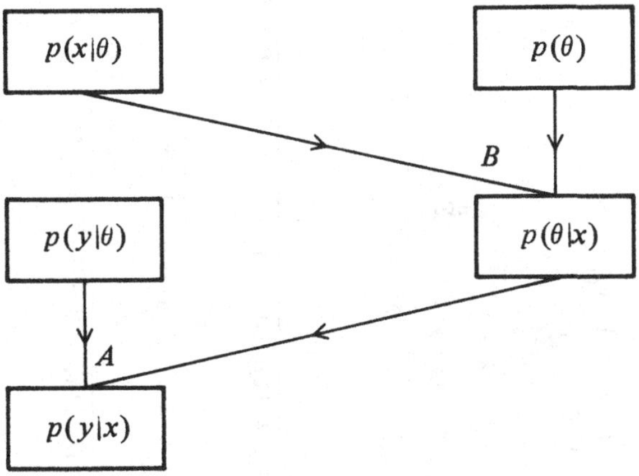

Fig. 2.1 Basic steps leading to the predictive density function.

Fig. 2.1 shows the basic steps leading to the predictive density function. The mechanism for arriving at node B, where the lines from $p(\theta)$ and $p(x\,|\,\theta)$ converge, is an operation of Bayes's theorem by (2.2) and (2.3). The subsequent process for obtaining the predictive density function at node A is an accumulation (2.4): the $p(y\,|\,\theta)$ and $p(\theta\,|\,x)$ on the lines leading to A are multiplied and accumulated over the common symbol θ, that is, either integrated or summed over Θ. The notation to be used for the more commonly encountered predictive distributions is shown in table 2.2. These distributions are also included in the complete list of notation in appendix I.

Table 2.2 *Predictive density functions*

Notation and name	Random variable and domain	Probability density function	Parameter restrictions
1. BeBi(n, g, h) Beta-binomial	y $y = 0, 1, \ldots, n$	$\binom{n}{y} \dfrac{B(g+y, h+n-y)}{B(g,h)}$	n positive integer, $g > 0, h > 0$.
2. NeBi(n, l) Negative-binomial	y $y = 0, 1, 2, \ldots$	$\binom{y+n-1}{n-1} l^y (1-l)^n$	n positive integer, $0 \leqslant l \leqslant 1$.
3. InBe(k, g, h) Inverse-beta	y $y > 0$	$\dfrac{h^g y^{k-1}}{B(k,g)(h+y)^{g+k}}$	$k > 0, g > 0, h > 0$.
4. DiMu$(n, \mathbf{g}, h)$ Dirichlet-multinomial	$\mathbf{y} = (y_1, y_2, \ldots, y_d)$ $y_i = 0, 1, \ldots, n$ $\Sigma y_i \leqslant n$	$\binom{n}{\mathbf{y}} \dfrac{D(\mathbf{g}+\mathbf{y}, h+n-\sum_1^d y_i)}{D(\mathbf{g}, h)}$	$\mathbf{g} = (g_1, \ldots, g_d)$, $g_i \geqslant 0$, $h > 0$.
5. St(k, b, c) Student	m $m \in \mathbf{R}^1$	$\dfrac{1}{B(\frac{1}{2}, \frac{1}{2}k)(kc)^{1/2}\{1 + (kc)^{-1}(m-b)^2\}^{(k+1)/2}}$	$b \in \mathbf{R}^1$, $c > 0, k > 0$.
Si(k, g, h) Siegel	v $v > 0$	$\dfrac{v^{(g/2)-1}}{B(\frac{1}{2}k, \frac{1}{2}g) h^{g/2} (1 + h^{-1}v)^{(k+g)/2}}$	$k > 0$, $g > 0, h > 0$.
StSi$(k; b, c; g, h)$ Student-Siegel	(m, v) $m \in \mathbf{R}^1, v > 0$	$\dfrac{v^{(g/2)-1}}{D(\frac{1}{2}, \frac{1}{2}k, \frac{1}{2}g)(kc)^{1/2} h^{g/2}\{1 + (kc)^{-1}(m-b)^2 + h^{-1}v\}^{(k+g+1)/2}}$	$b \in \mathbf{R}^1, c > 0$, $k > 0, g > 0, h > 0$.

Table 2.2 (*continued*)

Notation and name	Random variable and domain	Probability density function	Parameter restrictions
6. $\mathrm{St}_d(k, \mathbf{b}, \mathbf{c})$ Student	$\mathbf{m}$ $\mathbf{m} \in \mathbf{R}^d$	$\dfrac{1}{D\{\frac{1}{2}\mathbf{1}, \frac{1}{2}(k-d+1)\}\lvert k\mathbf{c}\rvert^{1/2}\{1+(\mathbf{m}-\mathbf{b})'(k\mathbf{c})^{-1}(\mathbf{m}-\mathbf{b})\}^{(k+1)/2}}$	$\mathbf{b} \in \mathbf{R}^d$, $\mathbf{c} \in \mathcal{S}^d$, $k > d-1$.
$\mathrm{Si}_d(k, g, \mathbf{h})$ Siegel	$\mathbf{v}$ $\mathbf{v} \in \mathcal{S}^d$	$\dfrac{\lvert\mathbf{v}\rvert^{(g-d-1)/2}}{B_d(\frac{1}{2}k, \frac{1}{2}g)\lvert\mathbf{h}\rvert^{g/2}\lvert\mathbf{I}+\mathbf{h}^{-1}\mathbf{v}\rvert^{(k+g)/2}}$	$k > d-1, g > d-1$, $\mathbf{h} \in \mathcal{S}^d$.
$\mathrm{StSi}_d(k; \mathbf{b}, \mathbf{c}; g, \mathbf{h})$ Student-Siegel	$(\mathbf{m}, \mathbf{v})$ $\mathbf{m} \in \mathbf{R}^d$, $\mathbf{v} \in \mathcal{S}^d$	$\dfrac{\Gamma_d\{\frac{1}{2}(k+g+1)\}\lvert\mathbf{v}\rvert^{(g-d-1)/2}}{\Gamma_d(\frac{1}{2}k)\Gamma_d(\frac{1}{2}g)\pi^{d/2}\lvert k\mathbf{c}\rvert^{1/2}\lvert\mathbf{h}\rvert^{g/2}\lvert\mathbf{I}+(k\mathbf{c})^{-1}(\mathbf{m}-\mathbf{b})(\mathbf{m}-\mathbf{b})'+\mathbf{h}^{-1}\mathbf{v}\rvert^{(k+g+1)/2}}$	$k > d-1$, $\mathbf{b} \in \mathbf{R}^d$, $\mathbf{c} \in \mathcal{S}^d$, $g > d-1$, $\mathbf{h} \in \mathcal{S}^d$.

Example 2.1

Prediction for the gamma distribution. As an illustration we take the reader through the formal stages of the derivation of the predictive density function for the gamma family. Here e is described by

$$p(x|\theta) = \text{Ga}(k, \theta) \tag{2.6}$$

$$= \frac{\theta^k x^{k-1} \exp(-\theta x)}{\Gamma(k)} \quad (x > 0).$$

The conjugate prior distribution for θ is also of gamma type, namely

$$p(\theta) = \text{Ga}(g, h) \tag{2.7}$$

$$= \frac{h^g \theta^{g-1} \exp(-h\theta)}{\Gamma(g)} \quad (\theta > 0).$$

For the case of a single $\text{Ga}(k, \theta)$ observation x a straightforward application of Bayes's theorem yields

$$p(\theta|x) = \text{Ga}(G, H) \text{ where } G = g + k, H = h + x. \tag{2.8}$$

The extension to the case where e consists of n replicates of a $\text{Ga}(k, \theta)$ experiment with observation $\mathbf{x} = (x_1, x_2, ..., x_n)$ is straightforward since Σx_i is sufficient for θ with probability distribution $\text{Ga}(nk, \theta)$. Hence

$$p(\theta|\mathbf{x}) = \text{Ga}(G, H) \text{ where } G = g + nk, H = h + \Sigma x_i.$$

To cover this possibility we take

$$p(y|\theta) = \text{Ga}(K, \theta) \tag{2.9}$$

where K is not necessarily equal to k. It follows that the predictive density function is given by

$$p(y|x) = \int_0^\infty \frac{\theta^K y^{K-1} \exp(-\theta y)}{\Gamma(K)} \frac{H^G \theta^{G-1} \exp(-H\theta)}{\Gamma(G)} d\theta$$

$$= \frac{H^G y^{K-1}}{\text{B}(K, G)(H + y)^{G+K}} \quad (y > 0),$$

so that

$$p(y|x) = \text{InBe}(K, G, H). \tag{2.10}$$

2.4 Predictive distributions for the standard situations

It is convenient to have available for easy reference a means of deriving such predictive distributions for standard situations. Table 2.3 is set out in a form corresponding to the derivation scheme of fig. 2.1, with the density functions in the following array:

$$\begin{array}{ll} p(x\,|\,\theta) & p(\theta) \\ p(y\,|\,\theta) & p(\theta\,|\,x) \\ p(y\,|\,x) & \end{array}$$

For instance, the five steps (2.6) to (2.10) of example 2.1 can be read immediately from the five entries of the array

$$\begin{array}{ll} \text{Ga}(k, \theta) & \text{Ga}(g, h) \\ \text{Ga}(K, \theta) & \text{Ga}(G, H) \\ \text{InBe}(K, G, H) & \end{array}$$

in section 3 of the table.

Although cases 1–3 of table 2.3 apparently deal with a single observation x from the informative experiment they are easily adapted to information consisting of a set $\mathbf{x} = (x_1, ..., x_n)$ of observations from n replicates. We have already seen an instance of this adaptation for the gamma case in example 2.1, and for the binomial, Poisson and gamma distributions the form of cases 1, 2 and 3 in table 2.3 can be directly used. For each of these distributions θ is one-dimensional and all the information about θ provided by the informative experiment is contained in the value of the sufficient statistic Σx_i which itself has a density function of the given form. For example, if each x_i is $\text{Po}(\theta)$ then the sufficient statistic Σx_i is $\text{Po}(n\theta)$. Thus from the table the posterior plausibility function $p(\theta\,|\,\mathbf{x})$ is simply $\text{Ga}(g + \Sigma x_i, h + n)$ and we can proceed to the construction of the predictive density function.

For the normal and multinormal distributions (cases 5 and 6 of table 2.3) the results are presented in a general way to allow for wide applicability. To motivate this form of presentation we consider the following situation. Suppose that the informative experiment consists of n replicates of a $\text{No}(\mu, \tau)$ experiment with outcomes $x_1, ..., x_n$. Here $\theta = (\mu, \tau)$ is two-dimensional and the data can be reduced to a sufficient statistic (m, v), where

$$m = \bar{x} = \sum x_i/n, \quad v = \sum (x_i - \bar{x})^2.$$

We can thus envisage the informative experiment as producing simply (m, v); moreover

$$p(m, v\,|\,\theta) = p(m\,|\,\mu, \tau)\,p(v\,|\,\tau),$$

so that, for given (μ, τ), m and v are independent with distributions of the following forms:

Table 2.3 *Construction of predictive distributions*

$p(x\mid\theta)$ $p(y\mid\theta)$ $p(y\mid x)$	$p(\theta)$ $p(\theta\mid x)$	
1 Binomial		
$\mathrm{Bi}(n, \theta)$ $\mathrm{Bi}(N, \theta)$ $\mathrm{BeBi}(N, G, H)$	$\mathrm{Be}(g, h)$ $\mathrm{Be}(G, H)$	where $G = g + x$, $H = h + n - x$.
2 Poisson		
$\mathrm{Po}(k\theta)$ $\mathrm{Po}(K\theta)$ $\mathrm{NeBi}\left(G, \dfrac{K}{K+H}\right)$	$\mathrm{Ga}(g, h)$ $\mathrm{Ga}(G, H)$	where $G = g + x$, $H = h + k$.
3 Gamma		
$\mathrm{Ga}(k, \theta)$ $\mathrm{Ga}(K, \theta)$ $\mathrm{InBe}(K, G, H)$	$\mathrm{Ga}(g, h)$ $\mathrm{Ga}(G, H)$	where $G = g + k$, $H = h + x$.
4 Multinomial		
$\mathrm{Mu}(n, \boldsymbol{\theta})$ $\mathrm{Mu}(N, \boldsymbol{\theta})$ $\mathrm{DiMu}(N, \mathbf{G}, H)$	$\mathrm{Di}(\mathbf{g}, h)$ $\mathrm{Di}(\mathbf{G}, H)$	where $\mathbf{G} = \mathbf{g} + \mathbf{x}$, $H = h + n - \mathbf{1}'\mathbf{x}$.

5 Normal

$$\begin{cases} \mathrm{No}(\mu, k\tau) \\ \mathrm{Ch}(\nu, \tau) \end{cases} \qquad \mathrm{NoCh}(b, c, g, h)$$

$$\begin{cases} \mathrm{No}(\mu, K\tau) \\ \mathrm{Ch}(\lambda, \tau) \end{cases} \qquad \mathrm{NoCh}(B, C, G, H) \quad \text{where } \begin{aligned} B &= C^{-1}(cb + km), \\ C &= c + k, \\ G &= g + \nu + \Delta(c), \\ H &= h + v + \frac{ck}{c+k}(m - b)^2 . \end{aligned}$$

$$\Delta(c) = \begin{cases} 0 & (c = 0), \\ 1 & (c > 0). \end{cases}$$

$$\begin{cases} \mathrm{St}\left\{G, B, \left(\frac{1}{K} + \frac{1}{C}\right) \frac{H}{G}\right\} \\ \mathrm{Si}(G, \lambda, H) \\ \mathrm{StSi}\left\{G; B, \left(\frac{1}{K} + \frac{1}{C}\right) \frac{H}{G}; \lambda, H\right\} \end{cases}$$

6 Multinormal

$$\begin{cases} \mathrm{No}_d(\boldsymbol{\mu}, k\boldsymbol{\tau}) \\ \mathrm{Wi}_d(\nu, \boldsymbol{\tau}) \end{cases} \qquad \mathrm{NoWi}_d(\mathbf{b}, c, g, \mathbf{h})$$

$$\begin{cases} \mathrm{No}_d(\boldsymbol{\mu}, K\boldsymbol{\tau}) \\ \mathrm{Wi}_d(\lambda, \boldsymbol{\tau}) \end{cases} \qquad \mathrm{NoWi}_d(\mathbf{B}, C, G, \mathbf{H}) \quad \text{where } \begin{aligned} \mathbf{B} &= C^{-1}(c\mathbf{b} + k\mathbf{m}), \\ C &= c + k, \\ G &= g + \nu + \Delta(c), \\ \mathbf{H} &= \mathbf{h} + \mathbf{v} + \frac{kc}{k+c}(\mathbf{m} - \mathbf{b})(\mathbf{m} - \mathbf{b})'. \end{aligned}$$

$$\begin{cases} \mathrm{St}_d\left\{G, \mathbf{B}, \left(\frac{1}{K} + \frac{1}{C}\right) \frac{\mathbf{H}}{G}\right\} \\ \mathrm{Si}_d(G, \lambda, \mathbf{H}) \\ \mathrm{StSi}_d\left\{G; \mathbf{B}, \left(\frac{1}{K} + \frac{1}{C}\right) \frac{\mathbf{H}}{G}; \lambda, \mathbf{H}\right\} \end{cases}$$

Table 2.3 (*continued*)

$p(x\mid\theta)$ $p(y\mid\theta)$ $p(y\mid x)$	$p(\theta)$ $p(\theta\mid x)$
7 Exponential (two-parameter)	
$\begin{cases}\mathrm{Er}(\mu, k\tau)\\ \mathrm{Ga}(\nu,\tau)\end{cases}$	$\mathrm{ElGa}(b, c, g, h)$
$\begin{cases}\mathrm{Er}(\mu, K\tau)\\ \mathrm{Ga}(\lambda,\tau)\end{cases}$	$\mathrm{ElGa}(B, C, G, H)$ where $B = \min(m, b)$, $C = c + k$, $G = g + \nu + \Delta(c)$, $H = h + \nu + \omega_m(b,c,k)(m-b)$. $\omega_m(b,c,k) = \begin{cases}-c & (m < b),\\ k & (m \geqslant b).\end{cases}$
$\begin{cases}\dfrac{CK}{\mathrm{B}(1, G)(C+K)H}\{1 + H^{-1}\omega_M(B, C, K)(M-B)\}^{-(G+1)}\\ \mathrm{InBe}(\lambda, G, H)\\ \dfrac{CKV^{\lambda-1}}{D(1, G, \lambda)(C+K)H^{\lambda+1}}\{1 + H^{-1}\omega_M(B, C, K)(M-B) + H^{-1}V\}^{-(G+\lambda+1)}\end{cases}$	

$$p(m|\mu,\tau) = \text{No}(\mu, k\tau),$$

$$p(v|\tau) = \text{Ch}(\nu, \tau),$$

where $k = n, \nu = n-1$. If the future experiment is to consist of N replicates of the $\text{No}(\mu, \tau)$ experiment we may be interested in such summary statistics as

$$M = \bar{y} = \sum y_i/N, \; V = \sum (y_i - \bar{y})^2,$$

which again are independent for given (μ, τ) and have distributions of the forms:

$$p(M|\mu,\tau) = \text{No}(\mu, K\tau),$$

$$p(V|\tau) = \text{Ch}(\lambda, \tau),$$

with $K = N, \lambda = N-1$. It is thus convenient to set out the predictive analysis in the following array:

$$p(x|\theta) \quad \begin{cases} p(m|\mu,\tau) \\ p(v|\tau) \end{cases} \qquad p(\theta)$$

$$p(y|\theta) \quad \begin{cases} p(M|\mu,\tau) \\ p(V|\tau) \end{cases} \qquad p(\theta|x)$$

$$p(y|x) \quad \begin{cases} p(M|x) \\ p(V|x) \\ p(M, V|x) \end{cases}$$

With such generality the results apply to even more complicated situations, as in the regression problem in §2.6. The result applies equally to simpler situations than that of the motivating example. For instance, if the informative experiment and the future experiment are each a single replicate of a $\text{No}(\mu, \tau)$ experiment then we simply apply case 5 of table 2.3 with $k = 1$, $\nu = 0, v = 0, K = 1$, to obtain $p(M|x)$ as the predictive density function of the outcome $y\,(= M)$ of the future experiment f.

For the multinormal case **m** and **M** are, of course, vectors and **v** and **V** are matrices but the concepts are completely analogous to those of the univariate normal case.

The idea underlying the presentation of the two-parameter exponential distribution (case 7 of table 2.3) is similar. For example, if the informative experiment e consists of n replicates of a $\text{Er}(\mu, \tau)$ experiment and we suppose that only the first r order statistics $x_{(1)}, \ldots, x_{(r)}$ $(2 \leqslant r \leqslant n)$ are available then

$$m = x_{(1)},$$

$$v = x_{(1)} + \ldots + x_{(r)} + (n-r)x_{(r)} - nx_{(1)}$$

are sufficient for (μ, τ), and are independent with

$$p(m \mid \mu, \tau) = \text{Er}(\mu, n\tau),$$
$$p(v \mid \tau) \quad = \text{Ga}(\nu, \tau),$$

where $\nu = r - 1$. Table 2.3 then provides the appropriate apparatus for the construction of the predictive density function for a $\text{Er}(\mu, K\tau)$ future experiment.

2.5 Prediction for regression models

A wide range of prediction problems occurs in regression situations; we have already seen a problem of this type in example 1.5. In such problems the future experiment f records the response (value of the dependent variable) made by some experimental unit to a known stimulus t (value of the explanatory or regressor variable), and so is better denoted by f_t than by f. We are thus led to consider a class

$$F = \{f_t : t \in T\}$$

of experiments, where each f_t has the same sample space Y and T is the set of possible stimuli. The dependence on t of the density function of f_t is also made explicit by the notation $p(y \mid t, \theta)$. Note that in specifying the density function for f_t we have used a common θ-index for every t rather than a possibly different index, say θ_t, for each t. We can achieve this by setting $\theta = \{\theta_t : t \in T\}$, for example, and we then gain some simplification in later developments, particularly in chapters 10 and 11.

The informative experiment e consists of a set of n independent experiments of given types $f_{t_1}, \ldots, f_{t_n}$ and provides a set of observations $(t_1, x_1), \ldots, (t_n, x_n)$, which can be portrayed as the usual scatter diagram in the (t, x) plane. To simplify notation we denote this complete set of observations by $\mathbf{z}$, and write $\mathbf{t} = (t_1, \ldots, t_n)$, $\mathbf{x} = (x_1, \ldots, x_n)$ when we wish to isolate the t and x components of $\mathbf{z}$.

At the centre of our analysis again will be the predictive density function for f_t, now denoted by $p(y \mid t, \mathbf{z})$. Corresponding to (2.5) we have

$$p(y \mid t, \mathbf{z}) = \int_\Theta p(y \mid t, \theta) p(\theta \mid \mathbf{z}) d\theta, \tag{2.11}$$

where

$$p(\theta \mid \mathbf{z}) = \frac{p(\theta) \prod_{i=1}^{n} p(x_i \mid t_i, \theta)}{\int_\Theta p(\theta) \prod_{i=1}^{n} p(x_i \mid t_i, \theta) d\theta}. \tag{2.12}$$

In accordance with our definition above of θ as a common index for all t we write $p(\theta\,|\,\mathbf{z})$ rather than $p(\theta\,|\,t, \mathbf{z})$ in (2.12). Despite the apparent complexity of (2.11) and (2.12) the derivation of the predictive regression density function can be easily obtained for standard situations by reference to the tables already presented in this chapter. We illustrate the technique for the binomial, Poisson, gamma and normal regression models.

(i) *Binomial.* Here f_t is $\mathrm{Bi}(t, \theta)$ and x, the total number of successes in f_{t_1}, $f_{t_2}, \ldots, f_{t_n}$, is sufficient for θ and is a $\mathrm{Bi}(\Sigma t_i, \theta)$ random variable. Hence given Σt_i we can use case 1 of table 2.3 with $n = \Sigma t_i$ and $N = t$ to derive $p(y\,|\,t, \mathbf{z})$.

(ii) *Poisson.* Here f_t is $\mathrm{Po}(t\theta)$ and, for given Σt_i, $x = \Sigma x_i$ is sufficient for θ with a $\mathrm{Po}(\Sigma t_i \theta)$ distribution.

(iii) *Gamma.* Similar remarks apply if f_t is $\mathrm{Ga}(t, \theta)$. Here, for given Σt_i, $x = \Sigma x_i$ is sufficient for θ with distribution $\mathrm{Ga}(\Sigma t_i, \theta)$.

(iv) *Normal.* The case of normal regression can also be analysed by means of table 2.3. Here f_t is described by a $\mathrm{No}(\alpha + \beta t, \tau)$ density function. Let

$$\bar{t} = \sum t_i/n, \; \bar{x} = \sum x_i/n,$$

$$S(\mathbf{t}, \mathbf{t}) = \sum (t_i - \bar{t})^2, \; S(\mathbf{t}, \mathbf{x}) = \sum (t_i - \bar{t})(x_i - \bar{x}),$$

$$S(\mathbf{x}, \mathbf{x}) = \sum (x_i - \bar{x})^2.$$

Then the estimated regression coefficient is

$$\hat{\beta} = S(\mathbf{t}, \mathbf{x})/S(\mathbf{t}, \mathbf{t})$$

and the residual sum of squares is

$$v = \sum \{x_i - \bar{x} - \hat{\beta}(t_i - \bar{t})\}^2 = S(\mathbf{x}, \mathbf{x}) - \frac{\{S(\mathbf{t}, \mathbf{x})\}^2}{S(\mathbf{t}, \mathbf{t})}.$$

Then, for given $S(\mathbf{t}, \mathbf{t})$,

$$\bar{x} + \hat{\beta}(t - \bar{t}) \text{ and } v$$

are jointly sufficient for

$$\mu = \alpha + \beta t \text{ and } \tau.$$

Also $\bar{x} + \hat{\beta}(t - \bar{t})$ and v are independently distributed as

$$\mathrm{No}\left[\mu, \tau \Big/ \left\{\frac{1}{n} + \frac{(t - \bar{t})^2}{S(\mathbf{t}, \mathbf{t})}\right\}\right] \text{ and } \mathrm{Ch}(n-2, \tau).$$

This is thus a special case of the normal array in case 5 of table 2.3 if we take

$$\frac{1}{k} = \frac{1}{n} + \frac{(t-\bar{t})^2}{S(t,t)} \quad \text{and} \quad \nu = n - 2.$$

Example 2.2

A bird-nesting problem. Four pairs of a rare species of bird nested for the first time in Scotland last season. The observed number of eggs in the four nests were 2, 3, 3, 4 and from these nests 1, 2, 3, 3 nestlings survived the season. At the start of the current season a new pair has a nest with 3 eggs. What are the chances that at least 2 nestlings will survive the season?

This is a case of binomial regression with θ representing the probability that an egg from a nest will give rise to a surviving nestling, with

t_i	2	3	3	4
x_i	1	2	3	3

and $t = 3$. Since the rare species was nesting under completely new conditions last season we use the vague prior distribution of table 2.1 with

$$p(\theta) = \text{Be}(g, h) \quad (g \to 0,\ h \to 0).$$

Hence, from (i) above, the predictive density function for y, the number of surviving nestlings in the new nest with $t = 3$ eggs, is given by table 2.3 as

$$p(y \mid t, \mathbf{z}) = \text{BeBi}(t, \Sigma x_i, \Sigma t_i - \Sigma x_i)$$
$$= \text{BeBi}(3, 9, 3).$$

We show the numerical values of this distribution in table 2.4, and so assess the chance that at least 2 nestlings will survive the season to be 0.371 + 0.453 = 0.824.

Table 2.4 *Predictive density function for number of surviving nestlings*

y	0	1	2	3
$p(y \mid t, \mathbf{z})$	0.027	0.148	0.371	0.453
$p\{y \mid t, \hat{\theta}(\mathbf{z})\}$	0.016	0.141	0.422	0.422

A technique of prediction commonly practiced is to regard the problem as primarily one of estimating θ, in this case by

$$\hat{\theta}(\mathbf{z}) = \sum x_i / \sum t_i = 0.75$$

say; and then to use the density function $p(y \mid t, \theta)$ with θ equal to $\hat{\theta}(\mathbf{z})$. This density function is shown in the last row of table 2.4 and is clearly quite different from $p(y \mid t, \mathbf{z})$. The difference arises from the fact that the last row uses $\hat{\theta}(z)$ as if it were the true index, whereas the construction of the predictive density function in (2.11) adopts the more reasonable approach of averaging the density functions $p(y \mid t, \theta)$, the weighting factors being

$$p(\theta \mid z) = \text{Be}(9, 3)$$

with mean value 0.75.

2.6 An application to medical prognosis

We conclude this chapter with a practical application which highlights the idea of predictive distributions. A recurring prediction problem in medical practice is that of prognosis. There the clinician in deciding which of the alternative available treatments is best for a particular patient must attempt to make prognoses of the future possible paths of the patient's illness on the different treatments. In any quantitative approach to such problems a first step is to provide for each treatment an appropriate predictive distribution for the future experiment which records the patient's progress. In the simple prognostic problem of example 1.1 a suitable measure of a patient's progress or effectiveness of treatment is taken to be survival time.

Table 2.5 *Ages and survival times of 20 carcinoma patients*

Age (years)	38	54	37	47	51	48	42	50	45	33
Survival time (weeks)	25	45	238	194	16	23	30	16	22	123
Age (years)	46	34	66	44	64	49	56	43	45	40
Survival time (weeks)	51	412	45	162	14	72	5	35	30	91

We assume that the 20 survival times of table 2.5 form a random sample from a lognormal distribution (Aitchison and Brown, 1957). By considering log (survival time) we can convert the data to a random sample $x_1, \ldots, x_n$ $(n = 20)$ from a $\text{No}(\mu, \tau)$ distribution. We then reduce the sample to the sufficient statistics

$$m = \sum x_i/n, \quad v = \sum (x_i - m)^2,$$

which are independently distributed as

$$\text{No}(\mu, n\tau), \quad \text{Ch}(n-1, \tau).$$

If we adopt a vague prior distribution for (μ, τ) as in case 5 of table 2.1 with

$$p(\mu, \tau) = \text{NoCh}(b, c, g, h) \quad (c \to 0, g \to 0, h \to 0)$$

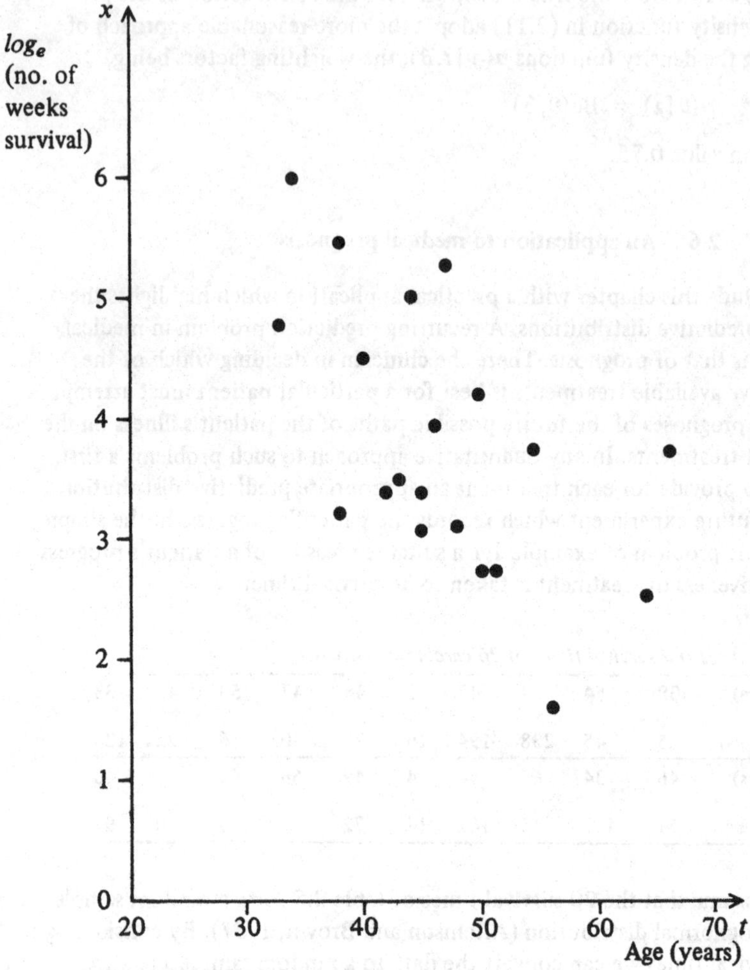

Fig. 2.2 Variability in survival time with age.

we have case 5 of table 2.3 with

$$k = n = 20,\ m = 3.83,\ \nu = n - 1 = 19,\ v = 23.22,\ K = 1.$$

We are led through the posterior distribution

$$p(\mu, \tau|\mathbf{x}) = \mathrm{NoCh}(m, n, \nu, v)$$

to the predictive density function

$$p(y\,|\,\mathrm{x}) = \mathrm{St}(19, 3.83, 1.28)$$

for y, the logarithm of survival time of a new patient on treatment of pre-operative radiotherapy followed by radical surgery.

The probability that a new patient will survive 150 weeks is then easily assessed by (A26) of appendix I as

$$\int_{5\cdot01}^{\infty} p(y\,|\,\mathrm{x})\mathrm{d}y = 0.5\, I_{0.946}(9.5, 0.5)$$

$$= 0.16.$$

Example 2.3

Age-specific medical prognosis. Table 2.5 shows the ages of the 20 carcinoma patients just considered. We now pose the question: to what extent is the reliability of prognosis improved by knowledge of this factor?

Fig. 2.2 shows a scatter diagram of age t and logarithm (x) of survival time and we recognise the possibility of explaining part of the variability of survival time in terms of a linear regression of x on t. Following the standard normal regression calculations of §2.5 we compute the basic quantities

$$n = 20, \quad \bar{t} = 46.6, \quad \bar{x} = 3.83,$$

$$S(\mathrm{t},\mathrm{t}) = 1460.8, \quad S(\mathrm{t},\mathrm{x}) = -104.25, \quad S(\mathrm{x},\mathrm{x}) = 23.22,$$

from which we obtain

$$\nu = 18, \quad \hat{\beta} = -0.0714, \quad v = 15.78.$$

A standard t-test shows that the regression coefficient is significantly different from zero at the 1 per cent level.

The predictive density function for y, the logarithm of the survival time of a new patient aged t and placed on the specified treatment, is

$$p(y\,|\,t,\mathrm{z}) = \mathrm{St}\{18, 7.16 - 0.0714t, 0.9205 + 0.000600(t - 46.6)^2\}.$$

The fact that the third parameter of this age-adjusted distribution is, for all t in the range of the informative experiment, less than the corresponding value 1.28 of the unadjusted predictive distribution is an indication of the greater precision of the age-adjusted analysis.

The age-specific nature of the survival rate can be presented in terms of probability curves for different survival times. Fig. 2.3 shows these curves for 100, 150 and 200 weeks and illustrates clearly how survival probability decreases with increasing age.

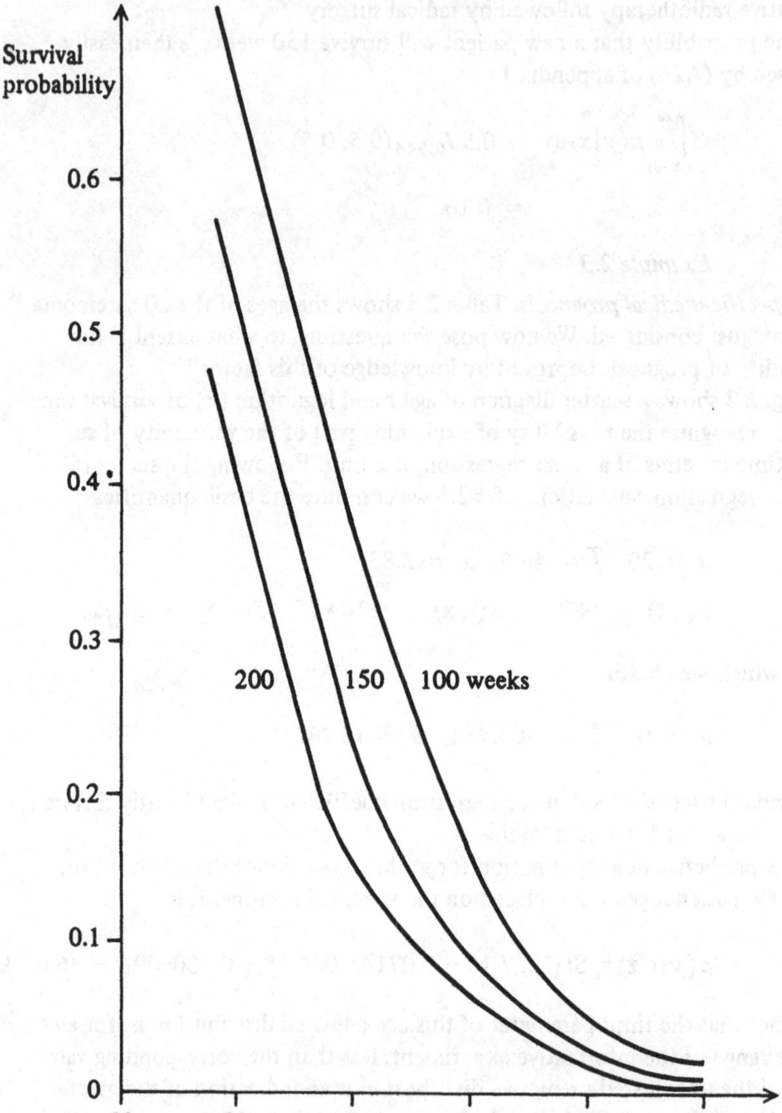

Fig. 2.3 Age-specific probability curves for different survival times.

History

The concept of the predictive distribution, at least in particular simple situations, has a long tradition dating back at least to Laplace. His 'rule of succession' was an attempt to answer the question: if in n binomial trials success has occurred n times what is the probability of success in the $(n + 1)$th trial? This question and its generalisation 'If we obtain x successes in n binomial trials what is the probability of y successes in the next N trials?' have often been the subject of discussion by philosophers and statisticians; for some comments on the history of this particular problem see the contribution by E.S. Pearson to the discussion of Aitchison (1964) and Thatcher (1964). Fisher (1935) provided a fiducial type derivation for a predictive distribution.

The widening of the concept and in particular its more extended use as an effective tool of data analysis are however fairly recent. Jeffreys (1961, p. 143) obtains the predictive distributions for the mean and sample variance of a second sample from a normal distribution based on the observations from a first sample and a vague prior distribution on the parameter. Geisser (1964) applies this to a diagnostic problem and Guttman and Tiao (1964) use predictive distributions for the normal and two-parameter exponential families. Zellner and Chetty (1965) derive the predictive distributions for the multivariate regression model. All these papers use only vague prior distributions. Aitchison and Sculthorpe (1965) analyse cases 1, 2, 3 and 5 of table 2.3 with general prior distributions. Raiffa and Schlaifer (1961, chapters 9–13) contains most of the relevant distribution theory of table 2.3 for the standard families. The idea of predictive distributions is not discussed by them but the derivation of

$$p(y) = \int_{\Theta} p(y|\theta)p(\theta)\mathrm{d}\theta$$

is given.

Further references: Ando and Kaufman (1965), Bratcher, Schucany and Hunt (1971), Chew (1966), de Finetti (1964), Geisser (1971), Kabe (1967), Lindley (1965*b*, pp. 11, 212), Lindley (1971, pp. 56–59), Roberts (1965), Winkler (1972, pp. 113, 204).

Problems

2.1 For the class of Pareto density functions

$$p(x|\theta) = \mathrm{Pa}(k,\theta) = \frac{\theta k^{\theta}}{x^{\theta+1}} \quad (x > k)$$

find a suitable conjugate class of prior density functions on the parameter set $\Theta = \{\theta : \theta > 0\}$.

Suppose that $\mathbf{x} = (x_1, ..., x_n)$ is a set of n independent observations from the above distribution. Obtain the predictive density function $p(y|\mathbf{x})$ based on this information and on a typical member of the conjugate class of prior density functions. Show that the probability that a future outcome exceeds ke^d, where $d > 0$, can be expressed in the form

$$\left[\frac{H}{H+d}\right]^G.$$

2.2 An informative experiment consists of n replicates yielding random variables $x_1, ..., x_n$, each distributed uniformly over the interval $(-\theta, \theta)$, where $\theta > 0$. The prior density function $p(\theta)$ is $\mathrm{Pa}(g, h)$. Show that the predictive distribution for a future observation y which is also, for given θ, uniformly distributed in $(-\theta, \theta)$ has density function

$$\frac{H}{2G(H+1)} \quad (|y| \leqslant G),$$

$$\frac{HG^H}{2(H+1)|y|^{H+1}} \quad (|y| > G),$$

where

$$G = \max\{g, |x_1|, ..., |x_n|\}, \quad H = h + n.$$

2.3 Show that if

$$p(x|\theta) = \frac{1}{\pi\{1+(x-\theta)^2\}} \quad (x \in R^1),$$

$$p(y|\theta) = \frac{1}{\pi\{1+(y-\theta)^2\}} \quad (y \in R^1),$$

and if the prior distribution is diffusely uniform over $\Theta = R^1$, then

$$p(y|x) = \frac{1}{2\pi\{1+\frac{1}{4}(y-x)^2\}} \quad (y \in R^1).$$

2.4 Show that, for an informative experiment with

$$p(x|\theta) = \tfrac{1}{2}\exp(-|x-\theta|) \quad (x \in R^1),$$

for a future experiment with

$$p(y|\theta) = \tfrac{1}{2}\exp(-|y-\theta|) \quad (y \in R^1),$$

and for a prior which is uniformly diffuse over the real line parameter space, the predictive density function is given by

$$p(y|x) = \tfrac{1}{4}(1+|y-x|)\exp(-|y-x|) \quad (y \in R^1).$$

2.5 A completely new extrusion process for the manufacture of artificial fibre is under investigation. It is assumed that the distribution of flaws along the length of the fibre follows a Poisson process. The numbers of flaws in five fibres of lengths 10, 15, 25, 30 and 40 metres were found to be 3, 2, 7, 6, 10 respectively. What is the probability that a fibre of length 60 metres will contain at most 14 flaws?

2.6 A machine uses N components and fails as soon as one or other of its components fails. The lifetimes of components manufactured during a long production run are known to be independently distributed as $\mathrm{Ex}(\theta)$, and θ is assumed to vary from run to run according to a $\mathrm{Ga}(g, h)$ distribution. From a recent production run life-testing of n components gave lifetimes $x_1, \ldots, x_n$. On the basis of all this information show that an appropriate distribution for the lifetime y of a machine constructed of components from this run has density function

$$\frac{NGH^G}{(H+Ny)^{G+1}} \quad (y > 0),$$

where

$$G = g+n, \; H = h + \sum_{i=1}^{n} x_i.$$

2.7 Suppose that independent identically distributed $\mathrm{Ex}(\theta)$ random variables $x_1, \ldots, x_n$ constitute the informative experiment and that the prior distribution for θ is $\mathrm{Ga}(g, h)$. Show that the predictive distribution for a future experiment described by independent identically distributed $\mathrm{Ex}(\theta)$ random variables $y_1, \ldots, y_N$ has density function

$$p(\mathbf{y}|\mathbf{x}) = \frac{\Gamma(G+N)}{\Gamma(G)} \frac{H^G}{(H+y_1+\ldots+y_N)^{G+N}} \quad (y_1 > 0, \ldots, y_N > 0),$$

where

$$G = g+n, \; H = h + \sum x_i.$$

Deduce the distribution of $z = \min(y_1, \ldots, y_N)$, and confirm that it is the same as the machine lifetime distribution of problem 2.6.

2.8 Consider a multinomial experiment with $N+1$ categories, of which N are 'success' categories with probabilities $\alpha_1, \ldots, \alpha_N$, the remaining failure category having probability $\alpha_{N+1} = 1 - \alpha_1 - \dots - \alpha_N$. Suppose that we record the numbers $y_1, \dots, y_N$ of the different types of success that occur before the kth failure is recorded. Then $y_1, \dots, y_N$ is said to have a negative multinomial distribution, written

$$\text{NeMu}(k; \alpha_1, \dots, \alpha_N, \alpha_{N+1}).$$

For a Poisson informative experiment with x distributed as $\text{Po}(t\theta)$, for a $\text{Ga}(g, h)$ prior distribution for θ, and for a future experiment yielding counts $y_1, \dots, y_N$ which, for given θ, are independent $\text{Po}(t_1\theta), \dots, \text{Po}(t_N\theta)$, show that the predictive joint distribution of $y_1, \dots, y_N$ is

$$\text{NeMu}\left\{g + x; \frac{t_1}{h + t + \Sigma t_i}, \dots, \frac{t_N}{h + t + \Sigma t_i}, \frac{h + t}{h + t + \Sigma t_i}\right\}.$$

Deduce the predictive distribution for the total count $y_1 + \dots + y_N$.

2.9 Each run of a process produces a large batch of ball bearings whose diameters (mm) are $\text{No}(\mu, \tau)$ distributed. The process is not sufficiently under control to achieve the target values $\mu = 8$, $\tau = 100$ on each run, and study of a large number of previous batches suggests that the variability of (μ, τ) from batch to batch follows a NoCh(8, 2, 1000, 10) distribution.

From a particular batch a sample of 10 ball bearings is chosen at random and their diameters (mm) are found to be

8.07, 8.15, 8.06, 7.97, 7.85, 8.02, 8.07, 8.17, 8.11, 8.09.

A further 15 bearings are selected at random from the batch. What are the probabilities
(i) that the mean diameter of the 15 bearings is less than 8.10;
(ii) that the standard deviation of the 15 diameters is less than 0.1?

2.10 The lifetimes (hours) of certain components are known to be independently distributed as $\text{Ex}(\tau)$. Ten of these components were simultaneously put into continuous use at an unrecorded time μ after their joint purchase and so far five of them have failed at times 2364, 2532, 2575, 2900, 3412 hours after purchase. An eleventh component from the same purchase is about to be put into continuous use. On the basis that your prior information about μ and τ is extremely imprecise (except, of course, that you know that $\mu > 0$) how do you assess the chances that this eleventh component will still be functioning after 2000 hours?

2.11 Suppose that the machine of problem 2.6 continues functioning until the Kth failure among its N components. Also suppose that the life testing of the n components from the recent production run is continued only until the kth component fails, and that these successive lifetimes are $x_1, \ldots, x_k$. Show that, given all this information, the probability that a machine survives at least time t is assessed as

$$N\binom{N-1}{K-1}\sum_{i=0}^{K-1}(-1)^i\binom{K-1}{i}\frac{1}{(N-K+i+1)}$$

$$\times\left\{\frac{h+v}{h+v+t(N-K+i+1)}\right\}^{g+k},$$

where $v = x_1 + \ldots + x_k + (n-k)x_k$.

2.12 To study the effectiveness of a fertiliser additive on the productivity of a certain variety of tomato plant, eight plants were grown in compost into which different strengths of the additive had been mixed. The resulting yields were as follows:

Strength of additive	1	2	3	4	5	6	7	8
Yield (kg)	1.28	1.84	1.37	1.50	2.20	1.65	3.22	3.00

On the basis of vague prior information concerning all the parameters involved how would you assess the probabilities
(i) that a single tomato plant grown in compost with additive at strength 4.5 will yield at least 2.5 kg;
(ii) that a single tomato plant grown in compost with additive at strength 8 will yield between 2.5 and 3.5 kg;
(iii) that eight tomato plants all grown in compost with additive at strength 5.5 will yield altogether more than 20 kg?

2.13 Let $x_1, \ldots, x_n$, associated with an informative experiment, be independent identically distributed random variables, each uniformly distributed over the unit interval $(\theta, \theta+1)$, and let y, associated with a future experiment, be also uniformly distributed over $(\theta, \theta+1)$. Suppose that the index set Θ is the interval (b, c) and that the prior distribution is the uniform distribution over (b, c). Write

$$m = \min\{x_1, \ldots, x_n\}, \; M = \max\{x_1, \ldots, x_n\}.$$

Show that, if the intervals (b, c) and $(M-1, m)$ intersect and if $B = \max\{b, M-1\}, C = \min\{c, m\}$, then

$$(C-B)p(y|\mathrm{x}) = \begin{cases} y-B & (B \leqslant y \leqslant C), \\ C-B & (C \leqslant y \leqslant B+1), \\ C+1-y & (B+1 \leqslant y \leqslant C+1). \end{cases}$$

What conclusions would you reach if the intervals (b, c) and $(M-1, m)$ were disjoint?

2.14 The two-parameter Pareto distribution, written $\mathrm{Pa}(\mu, \tau)$, has density function

$$\tau\mu^{\tau}/x^{\tau+1} \quad (x \geqslant \mu).$$

Suppose that $x_1, \ldots, x_n$ are independent $\mathrm{Pa}(\mu, \tau)$ random variables. Show that

$$m = \min(x_1, \ldots, x_n)$$

and

$$v = \log \frac{x_1 \ldots x_n}{m^n}$$

are sufficient statistics for (μ, τ) and are independently distributed as $\mathrm{Pa}(\mu, n\tau)$ and $\mathrm{Ga}(n-1, \tau)$.

For prior distribution of Pareto-Gamma type, written $\mathrm{PaGa}(b, c, g, h)$ and with density function

$$p(\mu, \tau) = \frac{c\tau\mu^{c\tau-1}}{b^{c\tau}} \frac{h^g \tau^{g-1} \mathrm{e}^{-h\tau}}{\Gamma(g)} \qquad (0 < \mu \leqslant b, \tau > 0),$$

show that the posterior distribution is $\mathrm{PaGa}(B, C, G, H)$, where

$$B = \min(b, m),$$

$$C = c + n,$$

$$G = g + n - 1 + \Delta(c),$$

$$H = h + v + \omega_m(b, c, n) \log(m/b).$$

Find the predictive distribution for a future observation y which, for given (μ, τ), is distributed as $\mathrm{Pa}(\mu, \tau)$.

3
Decisive prediction

3.1 Point prediction

If we are asked to predict the outcome of a performance of a future experiment f our answer will clearly depend on how we view the consequences of being wrong. More specifically we may attempt to assess the relative consequences of being 'close' to the realised outcome and of being 'badly' wrong. If we can quantify these visualised consequences then we can present the problem as one of statistical decision theory. Since in constructing the predictive density function $p(y|x)$ we have already carried out the information-extraction part of the problem we have a particularly simple confrontation in this decision problem. The components are as follows.

(i) *Parameter set*. The unknown outcome of the future experiment f plays the role of an unknown state of nature, so that Y, the sample space of f, is the parameter set of the statistical decision problem. Our assessment of the plausibility of a particular y at the time of making a decision is $p(y|x)$, the predictive density at y.

(ii) *Action set*. The set A of possible actions is simply a reproduction of Y, since any element of Y is a possible prediction a.

(iii) *Utility function*. Associated with each prediction or action a and each realisable outcome y there is a utility or value $U(a, y)$. We thus suppose defined a function U on the product domain $A \times Y$.

Standard statistical decision theory then directs us to choose as optimum a^* a prediction which maximises the expected utility

$$U(a) = \int_Y U(a, y) p(y|x) dy. \tag{3.1}$$

Thus

$$U(a^*) = \max_A U(a). \tag{3.2}$$

The optimum prediction depends on the particular x observed, and as we allow x to run through X we generate a function with domain X and range space A (= Y). Such a function, instructing as to what simple prediction or action a to adopt when faced with any outcome x of the informative experiment, may be termed a simple predictor.

Definition 3.1

Simple predictor. A *simple predictor* δ is a function

$$\delta : X \rightarrow A = Y. \tag{3.3}$$

What we have determined by the construction above is an optimum simple predictor δ^* defined by the optimum property

$$\max U\{\delta^*(x)\} = \max_A U(a). \tag{3.4}$$

Note that all we are doing here is to construct the Bayesian decision procedure relative to $p(y|x)$.

We now examine some simple results which depend only on the form of the utility function and not on the specific form of predictive density function. We confine attention to the one-dimensional versions of these results where Y is the real line or a subset of the real line.

All-or-nothing point prediction. How should a predictor be constructed if it is desperately important to predict the true outcome? For such a problem the natural formulation is to consider the limiting case ($\epsilon \rightarrow 0$) of the following utility specification:

$$U(a,y) = \begin{cases} 1 & (y-\epsilon < a < y+\epsilon), \\ 0 & \text{otherwise}. \end{cases} \tag{3.5}$$

Then, by (3.1),

$$\begin{aligned} U(a) &= \int_{a-\epsilon}^{a+\epsilon} p(y|x)dy \\ &= 2\epsilon\, p(a|x) + o(\epsilon) \quad (\epsilon \rightarrow 0), \end{aligned}$$

given simple regularity conditions on p. To maximise $U(a)$ we must maximise $p(a|x)$, and so we must adopt the intuitively reasonable procedure of predicting the most plausible outcome on the basis of the information we have. Here $\delta^*(x)$ is determined by

$$p\{\delta^*(x)|x\} = \max_A p(a|x). \tag{3.6}$$

In words: for an all-or-nothing utility structure the optimum simple prediction is the *mode* of the predictive distribution.

Linear loss point prediction. If the loss is zero when we predict correctly but is otherwise proportional to the distance of the prediction from the actual outcome then

$$U(a,y) = \begin{cases} -\xi(a-y) & (y<a), \\ -\eta(y-a) & (y \geqslant a), \end{cases} \tag{3.7}$$

where $\xi > 0, \eta > 0$. Then

$$U(a) = -\xi \int_{-\infty}^{a} (a-y)p(y\,|\,x)\mathrm{d}y - \eta \int_{a}^{\infty} (y-a)p(y\,|\,x)\mathrm{d}y$$

and the first and second derivatives of $U(a)$ with respect to a are

$$U'(a) = -\xi \int_{-\infty}^{a} p(y\,|\,x)\mathrm{d}y + \eta \int_{a}^{\infty} p(y\,|\,x)\mathrm{d}y,$$

$$U''(a) = -(\xi + \eta)p(a\,|\,x) < 0.$$

Hence $U(a)$ has a maximum where $U'(a) = 0$, so that the maximising a^* is given by

$$\int_{-\infty}^{a^*} p(y\,|\,x)\mathrm{d}y = \frac{\eta}{\xi + \eta}. \tag{3.8}$$

Thus for a linear loss utility structure the optimum simple prediction is the $\eta/(\xi + \eta)$-*quantile* of the predictive distribution. Only the relative value η/ξ is of importance and so, for example, we could assume that $\xi = 1$ without loss of generality. However it is convenient to retain both ξ and η in our formulation.

Note that if Y is one of the standard discrete spaces consisting of a set of integers we have to choose a^* such that

$$\sum_{0}^{a^*-1} p(y\,|\,x) \leqslant \frac{\eta}{\xi + \eta} \quad \text{and} \quad \sum_{0}^{a^*} p(y\,|\,x) > \frac{\eta}{\xi + \eta}. \tag{3.9}$$

We shall see later (§7.2) that this form of decisive prediction is related to a particular form of informative prediction.

Quadratic loss point prediction. If the loss is zero for a correct prediction and is proportional to the square of the error for a wrong prediction then we have

$$U(a,y) = -(a-y)^2 \quad (a \in A, y \in Y) \tag{3.10}$$

and

$$\begin{aligned} U(a) &= -\int_{Y} (a-y)^2 p(y\,|\,x)\mathrm{d}y \\ &= -\mathrm{V}(y\,|\,x) - \{a - \mathrm{E}(y\,|\,x)\}^2, \end{aligned}$$

where E and V denote the operations of evaluating the mean and variance of the indicated distribution. Hence $U(a)$ takes its largest value, $-\mathrm{V}(y\,|\,x)$, when $a = \mathrm{E}(y\,|\,x)$ and so we set

Table 3.1 *Characteristics of predictive distributions*

$p(y\|x)$	Mode	q-quantile $\left(q = \frac{\eta}{\xi + \eta}\right)$	Mean
BeBi(N, G, H)	$\begin{cases} 0 & \text{if } G < 1 \\ \left[\frac{(G-1)(N+1)}{G+H-2}\right] & \text{if } G \geqslant 1 \end{cases}$	largest integer a^* such that $P_{\mathrm{hy}}(N + G + H - 1, a^* + G - 1, N, a^* - 1) \leqslant q$ where P_{hy} is defined in (A21) of Appendix I	$\frac{NG}{G+H}$
NeBi$\left(G, \frac{K}{K+H}\right)$	$\begin{cases} 0 & \text{if } G < 1 \\ \left[\frac{(G-1)K}{H}\right] & \text{if } G \geqslant 1 \end{cases}$	largest integer a^* such that $I_{K/(K+H)}(a^*, G) \geqslant 1 - q$ where I is defined in (A19) of Appendix I	$\frac{GK}{H}$
InBe(K, G, H)	$\frac{H(K-1)}{G+1}$	solution a^* of $I_{a^*/(a^*+H)}(K, G) = q$ where I is defined in (A19) of Appendix I	$\frac{KH}{G-1}$
St$\left\{G, B, \left(\frac{1}{K} + \frac{1}{C}\right)\frac{H}{G}\right\}$	B	$B + \left\{\left(\frac{1}{K} + \frac{1}{C}\right)\frac{H}{G}\right\}^{1/2} \mathrm{t}(G; q)$ where t is defined in (A17) of Appendix I	B

Table 3.1 (*continued*)

$p(y \mid x)$	Mode	q-quantile $\left(q = \frac{\eta}{\xi + \eta}\right)$	Mean
$\frac{CKG}{(C+K)H}\left\{1 + \frac{\omega_y(B, C, K)}{H}(y - B)\right\}^{-(G+1)}$	B	$\begin{cases} B + \frac{H}{C} - \frac{H}{C}\left\{\frac{(C+K)q}{K}\right\}^{-1/G} & \text{if } q \leqslant \frac{K}{C+K}, \\ B - \frac{H}{K} + \frac{H}{K}\left\{\frac{(C+K)}{C}(1-q)\right\}^{-1/G} & \text{if } q > \frac{K}{C+K} \end{cases}$	$B + \frac{(C-K)H}{CK(G-1)}$

$$\delta^*(x) = \mathrm{E}(y|x). \tag{3.11}$$

In words: for a quadratic loss structure the optimum prediction is the *mean* of the predictive distribution. For a discrete predictive distribution we would select the a-value which is closest to $\mathrm{E}(y|x)$.

To help in the applications of these types of prediction we provide in table 3.1 the mode, q-quantile and mean for each of the one-dimensional predictive distributions of table 2.2 for which a simple formulation is possible

3.2 An application to machine tool replacement

In the preceding section we have selected some simple utility functions and obtained some general results about optimum prediction. In any real application, however, an appropriate utility function should be constructed to meet the particular circumstances of the problem. We shall now illustrate various aspects of such decisive prediction by determining an optimum inspection and replacement policy for the machine tool problem of example 1.2. In the present section we recognise three different types of policy as point prediction decisions and determine which particular predictions are optimum for the three types. In order to see the structure of the analysis we retain the symbols ξ, η, ζ of example 1.2 rather than use their specific values. Later in §3.4 we shall see that the three types of policy are special cases of a more general policy which can be represented as an interval prediction decision. There we shall use the specific values of ξ, η, ζ and the data to pinpoint the overall optimum policy.

> *Policy 1.* Send in the inspector at time a to replace the machine tool immediately.

This is tantamount to predicting that the machine will wear out at time a, and we have to decide which is the best a^* to select.

First we resolve the question of the predictive distribution. We make the reasonable assumption that the individual lifetimes are independently and exponentially distributed, say as $\mathrm{Ex}(\theta)$, so that the sum $x(= 1939)$ of the $n(= 24)$ lifetimes is a sufficient statistic for θ. We may thus take as the informative experiment density function

$$p(x|\theta) = \mathrm{Ga}(n, \theta).$$

Suppose that we adopt a conjugate prior distribution on Θ with density function

$$p(\theta) = \mathrm{Ga}(g, h).$$

The future experiment of interest consists in the recording of the lifetime y of a further machine tool and so we have

$$p(y|\theta) = \text{Ex}(\theta) = \text{Ga}(1,\theta).$$

Application of case 3 of table 2.3 yields the predictive density function

$$p(y|x) = \text{InBe}(1, G, H), \tag{3.12}$$

where $G = g + n, H = h + x$.

The utility function $U(a, y)$ is readily obtained. If $y \leq a$ the machine tool is worn out before the inspector arrives and time $a - y$ of production is lost, so that the utility is then $-\xi(a-y)$. If $y > a$ then the machine tool is still functioning, and therefore there is a scrapping loss of ζ. We have also lost a time $y - a$ of production and so the total utility is $-\zeta - \xi(y-a)$. Hence

$$U(a,y) = \begin{cases} -\xi(a-y) & (y \leq a), \\ -\zeta - \xi(y-a) & (y > a). \end{cases} \tag{3.13}$$

We then have immediately, by (3.1), that

$$U(a) = -\xi\int_0^a (a-y)p(y|x)\mathrm{d}y - \int_a^\infty \{\zeta + \xi(y-a)\}p(y|x)\mathrm{d}y,$$

so that

$$U'(a) = \zeta p(a|x) - \xi\int_0^a p(y|x)\mathrm{d}y + \xi\int_a^\infty p(y|x)\mathrm{d}y$$

$$= \zeta p(a|x) + 2\xi\int_a^\infty p(y|x)\mathrm{d}y - \xi$$

$$= \xi\left[\frac{\zeta G}{\xi H}\left(\frac{H}{H+a}\right)^{G+1} + 2\left(\frac{H}{H+a}\right)^G - 1\right].$$

To see whether

$$U'(a) = 0 \quad (a \geq 0)$$

has a solution we set

$$w = H/(H+a) \tag{3.14}$$

and consider whether the equation

$$f(w) = \frac{\zeta G}{\xi H}w^{G+1} + 2w^G - 1 = 0 \quad (0 < w \leq 1) \tag{3.15}$$

has a solution. Since $f(0) = -1, f(1) > 0$ and $f'(w) > 0 \quad (0 < w \leq 1)$ we see that there is in fact a unique solution, w^* say. Thus

$$a^* = \frac{1-w^*}{w^*}H \tag{3.16}$$

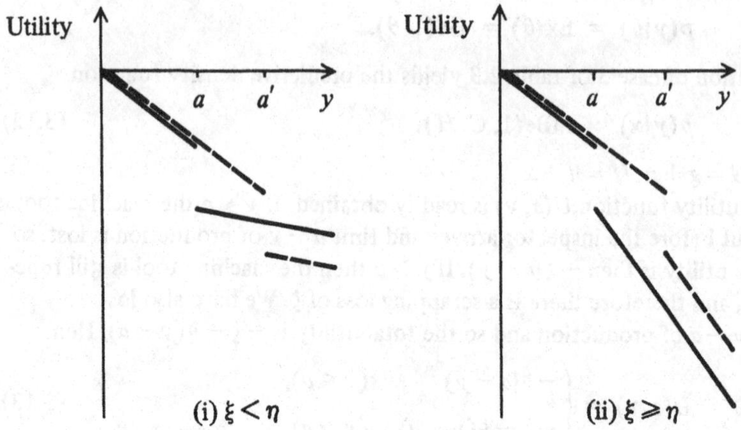

Fig. 3.1 Graphs of utility functions (3.21) for two predictions a, a' with $a < a'$.

———— $U(a, y)$
– – – – $U(a', y)$

provides the optimum prediction of the machine tool lifetime, or equivalently, the optimum time at which to replace.

After some simple integration and tedious algebra it can be shown that

$$U(a^*) = -\frac{\xi H}{G(G-1)w^*}\{2w^{*G} - G^2w^* + G^2 - 1\}. \tag{3.17}$$

We now consider an alternative policy.

> *Policy 2*. Send in the inspector at time a to replace the machine tool immediately if it has already worn out, and otherwise to attend the tool until it wears out.

For this policy the predictive distribution is as before but the utility function is now

$$U(a,y) = \begin{cases} -\xi(a-y) & (y<a), \\ -\eta(y-a) & (y \geqslant a). \end{cases} \tag{3.18}$$

(A loss of production is involved if the inspector arrives too late, a cost of labour if he arrives too early.) This is the linear loss utility structure (3.7) and so the optimum a^* is the $\eta/(\xi+\eta)$-quantile of the predictive distribution, satisfying

$$\int_0^{a^*} p(y|x)\,dy = \frac{\eta}{\xi+\eta}.$$

Direct integration and solving the resulting equation for a^* gives

$$a^* = H\left\{\left(\frac{\xi+\eta}{\xi}\right)^{1/G} - 1\right\}. \tag{3.19}$$

Again simple integration and some algebra yield the result

$$U(a^*) = -\frac{\xi G a^*}{G-1}. \tag{3.20}$$

Let us consider also a third policy.

Policy 3. Make the inspector attend the tool from its start until it wears out or for time a, whichever is the shorter. If the machine has not worn out by time a replace it then.

Here

$$U(a,y) = \begin{cases} -\eta y & (y \leqslant a), \\ -\zeta - \eta a - \xi(y-a) & (y > a). \end{cases} \tag{3.21}$$

We can immediately distinguish between two cases (i) $\xi < \eta$ and (ii) $\xi \geqslant \eta$. For each of these cases fig. 3.1 shows the graphs of $U(a,y)$ against y for two predictions a and a' with $a < a'$. In case (ii) $U(a,y) \leqslant U(a',y)$ for every y. For this case it follows that $a^* \to \infty$, as we might expect since $\xi \geqslant \eta$ means that the loss of production is greater than or equal to the cost of labour per unit time. We therefore restrict attention to the case

$$\xi < \eta \tag{3.22}$$

for which we shall obtain a non-trivial prediction.

The expected utility can here be expressed as

$$U(a) = -\eta \mathrm{E}(y\,|\,x) - \int_a^\infty \{\zeta + \eta(a-y) + \xi(y-a)\}p(y\,|\,x)\mathrm{d}y.$$

Hence

$$U'(a) = \zeta p(a\,|\,x) - (\eta - \xi)\int_a^\infty p(y\,|\,x)\mathrm{d}y$$

$$\begin{cases} = 0 \text{ at } a^* = \dfrac{\zeta G}{\eta-\xi} - H \quad \text{if} \quad \dfrac{\zeta G}{\eta-\xi} > H, \\ < 0 \text{ for all } a > 0 \text{ otherwise.} \end{cases}$$

The maximising a^* is thus given by

$$a^* = \begin{cases} \dfrac{\zeta G}{\eta-\xi} - H & \text{if} \quad \dfrac{\zeta G}{\eta-\xi} > H, \\ 0 & \text{if} \quad \dfrac{\zeta G}{\eta-\xi} \leqslant H. \end{cases} \tag{3.23}$$

(If the scrapping loss is small enough it pays to replace the machine tool immediately.) Again a simple expression for the maximum expected utility can be obtained:

$$U(a^*) = \begin{cases} -\dfrac{1}{G-1}\left\{\eta H - \zeta\left(\dfrac{H}{H+a^*}\right)^G\right\} & (a^* > 0), \\ -\zeta - \dfrac{\xi H}{G-1} & (a^* = 0). \end{cases} \tag{3.24}$$

3.3 Set prediction

Often there may be no great pressure to pinpoint the actual outcome of the future experiment f but rather a need to ensure that an interval, region or set is provided which contains the realised outcome. For the purposes of design, for example, it may not be so important that we know that a component functions at a specified value of some characteristic as to be fairly sure that its operation lies within some acceptable range. For such predictive problems we naturally take as our action set A not the set Y of possible outcomes but the class $\mathcal{Y}$ of (measurable) subsets of Y. Presumably if we specify a set prediction a we are happy if the actual outcome y falls in a, unhappy if y falls outside a and, if degrees of happiness are allowable, most unhappy when y falls well outside a. If we can quantify this then we again have a utility structure with utility $U(a, y)$ defined for every $a \in A$ and $y \in Y$.

We shall examine in some detail the case where Y is the whole or part of the real line and where a is restricted by the practical requirements of the problem to be an interval (a_1, a_2). As for simple prediction we choose an interval a which maximises the expected utility

$$U(a) = \int_Y U(a, y) p(y|x) dy.$$

The tensions in the construction of the utility function here will be on the one hand a desire to keep the interval short and so provide a useful prediction and on the other hand a fear of the losses involved if the interval fails to capture the outcome.

The definition of a simple predictor extends in an obvious way to this more general situation.

Definition 3.2

Set predictor. A *set predictor* δ is a function

$$\delta : X \to A = \mathcal{Y}. \tag{3.25}$$

The notion of optimality is again as defined in (3.2). It is possible that the optimum set prediction may reduce to a point prediction. As in §3.1 we can investigate the possibility of some general results, which depend on the special form of the utility function but not on the form of the predictive distribution.

All-or-nothing set prediction. Here we consider a utility structure which awards a unit if and only if the interval contains the outcome of f and where there is a cost proportional to the length of the interval. More precisely,

$$U(a_1, a_2, y) = \begin{cases} 1 - \gamma(a_2 - a_1) & (a_1 \leqslant y \leqslant a_2), \\ -\gamma(a_2 - a_1) & \text{otherwise.} \end{cases} \tag{3.26}$$

The expected utility is then

$$\begin{aligned} U(a_1, a_2) &= \int_{a_1}^{a_2} p(y|x)\mathrm{d}y - \gamma(a_2 - a_1) \\ &= \int_{a_1}^{a_2} \{p(y|x) - \gamma\}\mathrm{d}y. \end{aligned}$$

If therefore we place in $(a_1{}^*, a_2{}^*)$ all values of y such that $p(y|x) > \gamma$ we will maximise $U(a_1, a_2)$. Hence

$$(a_1{}^*, a_2{}^*) = \{y : p(y|x) > \gamma\}, \tag{3.27}$$

as illustrated in fig. 3.2.

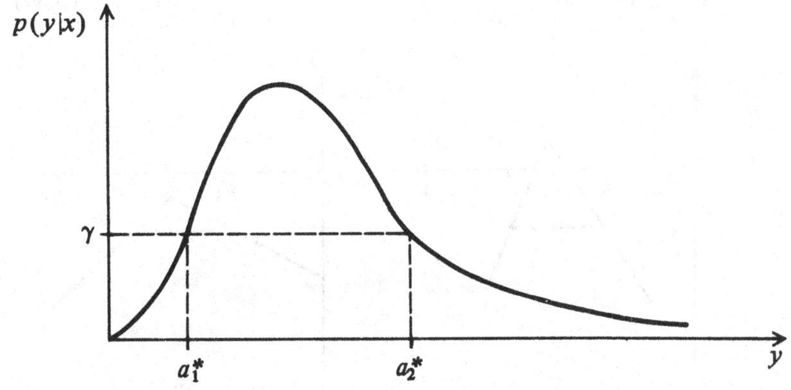

Fig. 3.2 Optimum set prediction for utility function (3.26).

If γ is of such a magnitude that no values of y satisfy $p(y|x) > \gamma$ then a limiting argument similar to that for the all-or-nothing point prediction of §3.1 shows that the best interval prediction degenerates into a point prediction, the mode of the predictive density function $p(y|x)$.

Linear utility interval prediction. Suppose that we assert that an interval (a_1, a_2) will contain the outcome y of the future experiment f. Suppose further that we incur penalties if in fact y lies below a_1 or above a_2, and these penalties are proportional to the amounts by which y escapes the interval. If the cost attaching to the interval used is proportional to the length of the interval, say $\gamma(a_2 - a_1)$, then we are led to investigate the piecewise-linear utility function

$$U(a_1, a_2, y) = \begin{cases} -\xi(a_1 - y) - \gamma(a_2 - a_1) & (y < a_1), \\ -\gamma(a_2 - a_1) & (a_1 \leqslant y \leqslant a_2), \\ -\eta(y - a_2) - \gamma(a_2 - a_1) & (y > a_2), \end{cases} \quad (3.28)$$

where ξ, η, γ are all positive constants. Since only relative values of ξ, η and γ are of importance we take $\gamma = 1$ without loss of generality. By varying ξ and η we allow for the possibility of differential losses associated with the interval overshooting or undershooting the true y.

If $1/\xi + 1/\eta \leqslant 1$ we find that maximisation of the expected utility leads to selecting an interval $(a_1{}^*, a_2{}^*)$ given by the solutions of

$$\int_{-\infty}^{a_1{}^*} p(y|x)\mathrm{d}y = 1/\xi, \quad \int_{a_2{}^*}^{\infty} p(y|x)\mathrm{d}y = 1/\eta. \quad (3.29)$$

If however $1/\xi + 1/\eta > 1$ the optimal interval prediction degenerates into a point prediction a^* given by the solution of

$$\int_{-\infty}^{a^*} p(y|x)\mathrm{d}y = \frac{\eta}{\xi + \eta}. \quad (3.30)$$

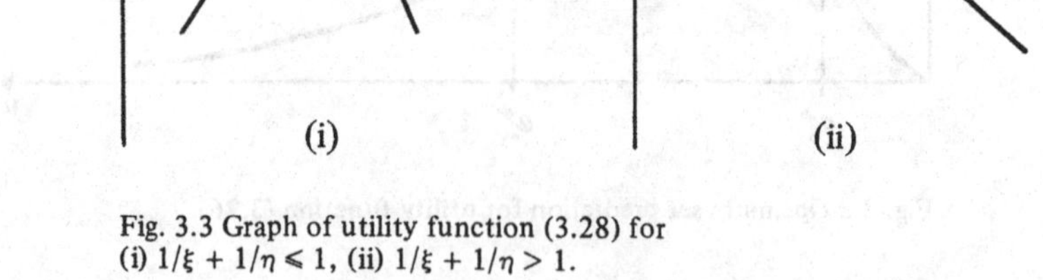

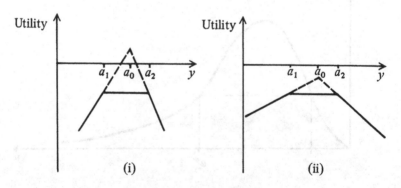

Fig. 3.3 Graph of utility function (3.28) for (i) $1/\xi + 1/\eta \leqslant 1$, (ii) $1/\xi + 1/\eta > 1$.

Fig. 3.3 shows the graph of the utility function (3.28) for the two cases of the inequality. For the case $1/\xi + 1/\eta > 1$ every interval prediction $a = (a_1, a_2)$

is inadmissible since there exists a point prediction, namely $a_0 = (\xi a_1 + \eta a_2)/(\xi + \eta)$, with uniformly larger expected utility.

3.4 Further analysis of the machine tool problem

As an example of set or interval prediction we can consider a more general policy than we have so far investigated.

> *Policy 4*. Send in the inspector at time a_1 and have him replace the machine tool as soon as it wears out or at time a_2 whichever is earlier.

One interpretation of such a policy is that we are regarding (a_1, a_2) as a prediction interval since it is clearly our hope that the lifetime y will fall in the interval (a_1, a_2). Note that the three policies previously considered are all special cases of this policy. For policy 1, $a_2 = a_1$; for policy 2, $a_2 = \infty$; for policy 3, $a_1 = 0$. It is thus of interest to ask under what conditions one of those three policies will turn out to be optimum within this more general policy framework.

For this case,

$$U(a_1, a_2, y) = \begin{cases} -\xi(a_1 - y) & (y < a_1), \\ -\eta(y - a_1) & (a_1 \leqslant y \leqslant a_2), \\ -\zeta - \eta(a_2 - a_1) - \xi(y - a_2) & (y > a_2). \end{cases} \tag{3.31}$$

For example if $y > a_2$ a scrapping loss ζ is involved, there is a cost of inspection for the whole interval, and we also have to debit for the unused lifetime $y - a_2$ of the machine tool. Again we assume that $\xi < \eta$. The expected utility can be expressed as the sum of a function of a_1 only and a function of a_2 only:

$$\begin{aligned} U(a_1, a_2) &= -\xi \int_0^{a_1} (a_1 - y) p(y \mid x) \mathrm{d}y - \eta \int_{a_1}^{a_2} (y - a_1) p(y \mid x) \mathrm{d}y \\ &\quad - \int_{a_2}^{\infty} \{\zeta + \eta(a_2 - a_1) + \xi(y - a_2)\} p(y \mid x) \mathrm{d}y \\ &= F_1(a_1) + F_2(a_2), \end{aligned}$$

where

$$F_1(a_1) = \eta a_1 - (\xi + \eta) \int_0^{a_1} (a_1 - y) p(y \mid x) \mathrm{d}y,$$

$$F_2(a_2) = -\eta \mathrm{E}(y \mid x) - \int_{a_2}^{\infty} \{\zeta - (\eta - \xi)(y - a_2)\} p(y \mid x) \mathrm{d}y.$$

We have that

$$\frac{\partial U}{\partial a_1} = \eta - (\xi + \eta) \int_0^{a_1} p(y|x)\,dy$$

$$= \eta - (\xi + \eta)\left\{1 - \left(\frac{H}{H + a_1}\right)^G\right\}$$

and

$$\frac{\partial U}{\partial a_2} = \zeta p(a_2|x) - (\eta - \xi) \int_{a_2}^{\infty} p(y|x)\,dy$$

$$= \frac{(\eta - \xi)H^G}{(H + a_2)^{G+1}}\left\{\frac{\zeta G}{\eta - \xi} - H - a_2\right\}.$$

Note first that the equations

$$\frac{\partial U}{\partial a_1} = \frac{\partial U}{\partial a_2} = 0 \quad (a_1, a_2 > 0)$$

have roots

$$a_1{}^* = H\left\{\left(\frac{\xi + \eta}{\xi}\right)^{1/G} - 1\right\}, \tag{3.32}$$

$$a_2{}^* = \frac{\zeta G}{\eta - \xi} - H, \tag{3.33}$$

provided

$$\frac{\zeta G}{\eta - \xi} > H; \tag{3.34}$$

see policy 3. From now on we shall suppose that this condition holds.

Then it can easily be shown that at $(a_1{}^*, a_2{}^*)$

$$\frac{\partial^2 U}{\partial a_1{}^2} < 0, \quad \frac{\partial^2 U}{\partial a_2{}^2} < 0, \quad \frac{\partial^2 U}{\partial a_1\, \partial a_2} = 0.$$

Hence, if $a_1{}^* < a_2{}^*$, U attains its maximum value within the region $0 < a_1 < a_2$ at the local maximum $(a_1{}^*, a_2{}^*)$; otherwise U has its maximum on the boundary $a_2 = a_1$.

Hence $(a_1{}^*, a_2{}^*)$ is the optimum interval if and only if

$$\frac{\zeta G}{(\eta - \xi)} > H\left(\frac{\xi + \eta}{\xi}\right)^{1/G} \quad (> H). \tag{3.35}$$

In this case the maximum utility is

$$U(a_1{}^*, a_2{}^*) = -\frac{1}{G - 1}\left\{\xi G a_1{}^* - \zeta\left(\frac{H}{H + a_2{}^*}\right)^G\right\}. \tag{3.36}$$

For the case

$$\frac{\zeta G}{(\eta - \xi)} \leqslant H \left(\frac{\xi + \eta}{\xi} \right)^{1/G} \tag{3.37}$$

there is no local maximum in the region $0 < a_1 < a_2$ but we know that the absolute maximum lies on the boundary $a_2 = a_1$. In this case we obtain a point estimate which is the solution of

$$\frac{\zeta G}{\xi H} \left(\frac{H}{H + a} \right)^{G+1} + 2 \left(\frac{H}{H + a} \right)^{G} - 1 = 0,$$

that is the same solution as for policy 1; see (3.15).

We are now in a position to examine the consequences of using the specific values $\xi = 1.8, \eta = 2.4, \zeta = 54$. If we assume that information is vague concerning θ before the evidence of the lifetimes of the 24 machine tools then by case 3 of table 2.1 we take $g = h = 0$, so that $G = n = 24, H = x = 1939$. We note that $\xi < \eta$, that the first condition in (3.23) is satisfied and that condition (3.35) is also satisfied. Straightforward calculations then lead to optimum policies and maximised expected utilities as shown in table 3.2. Since policies 1, 2 and 3 are special cases of policy 4 the optimum form of policy 4 must be overall best, but table 3.2 shows that the advantage over the optimum form of policy 2 is slight.

Table 3.2 *Optima for the four types of policy*

Policy	Optimum a^* or $(a_1{}^*, a_2{}^*)$	Maximised $U(a^*)$ or $U(a_1{}^*, a_2{}^*)$
1	70.6	−131.5
2	69.7	−130.9
3	221.0	−202.2
4	69.7, 221.0	−130.7

For the specific relative values of ξ, η and ζ the choice of optimum policy is delicately balanced as we can easily demonstrate by considering changes in the rate η of labour cost for fixed values of ξ and ζ. Fig. 3.4 shows the graphs of $a_1{}^*$ and $a_2{}^*$ and fig. 3.5 the graphs of the maximised expected utilities for each of the four policies, plotted against η in the neighbourhood of $\eta = 2.4$. As η increases from 2.4 the advantage of policy 4 over policy 2 increases, while the advantage over policy 1 decreases. For $\eta \geqslant 2.445$ policy 1 is overall best. As η decreases from 2.4 the advantage of policy 4 over policy 2 decreases as the appropriate inspection period for policy 4 expands due to the reduction of labour costs. For $\eta < 1.8$ condition (3.22) is not satisfied and policy 2 is the optimum form. Only when $\eta = 0$ is policy 3 worth considering.

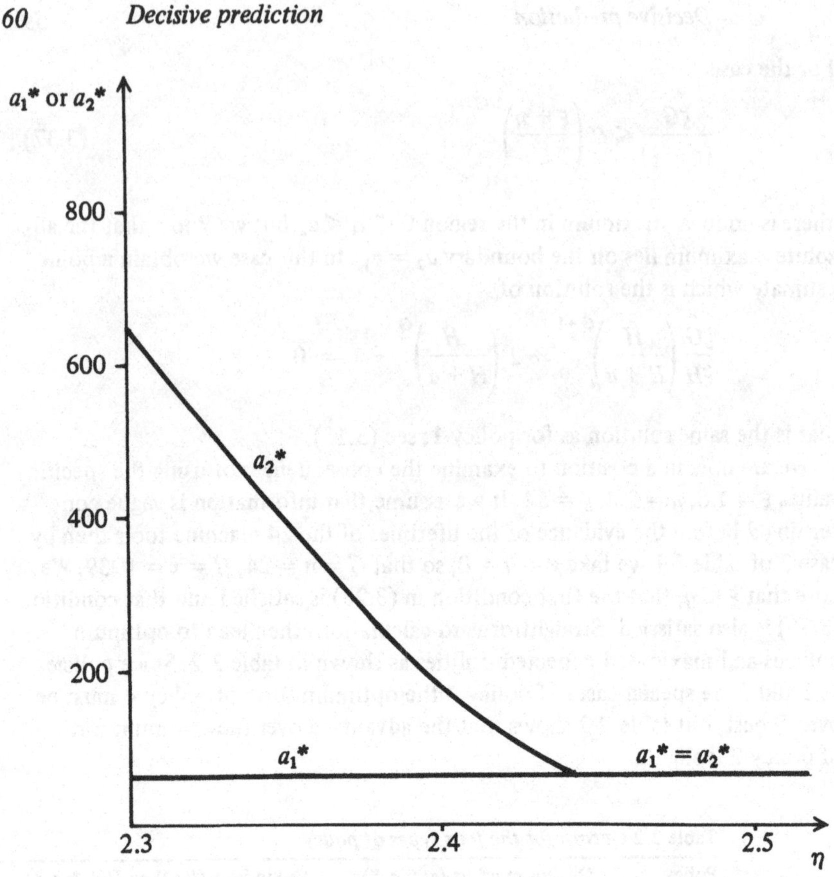

Fig. 3.4 Graphs of a_1* and a_2* against η for the machine tool problem.

3.5 Computational aids for more general utility functions

Earlier, in §§3.1 and 3.3, we obtained for some simple utility structures some general results applicable whatever the predictive distribution. The example we have just studied shows that for more complicated utility functions we are forced to take account of the particular form of $p(y|x)$ to obtain a useful result. If the utility function can be expressed as a polynomial in y, or is piecewise polynomial in y, then some help can be given in the evaluation of $U(a)$. For any polynomial or piecewise polynomial in y can be expressed as a linear combination of simpler utilities of one of the following forms:

$$U_j(a, y) = \begin{cases} y(y-1)\dots(y-j+1) & (y < a), \\ 0 & (y \geqslant a); \end{cases} \tag{3.38}$$

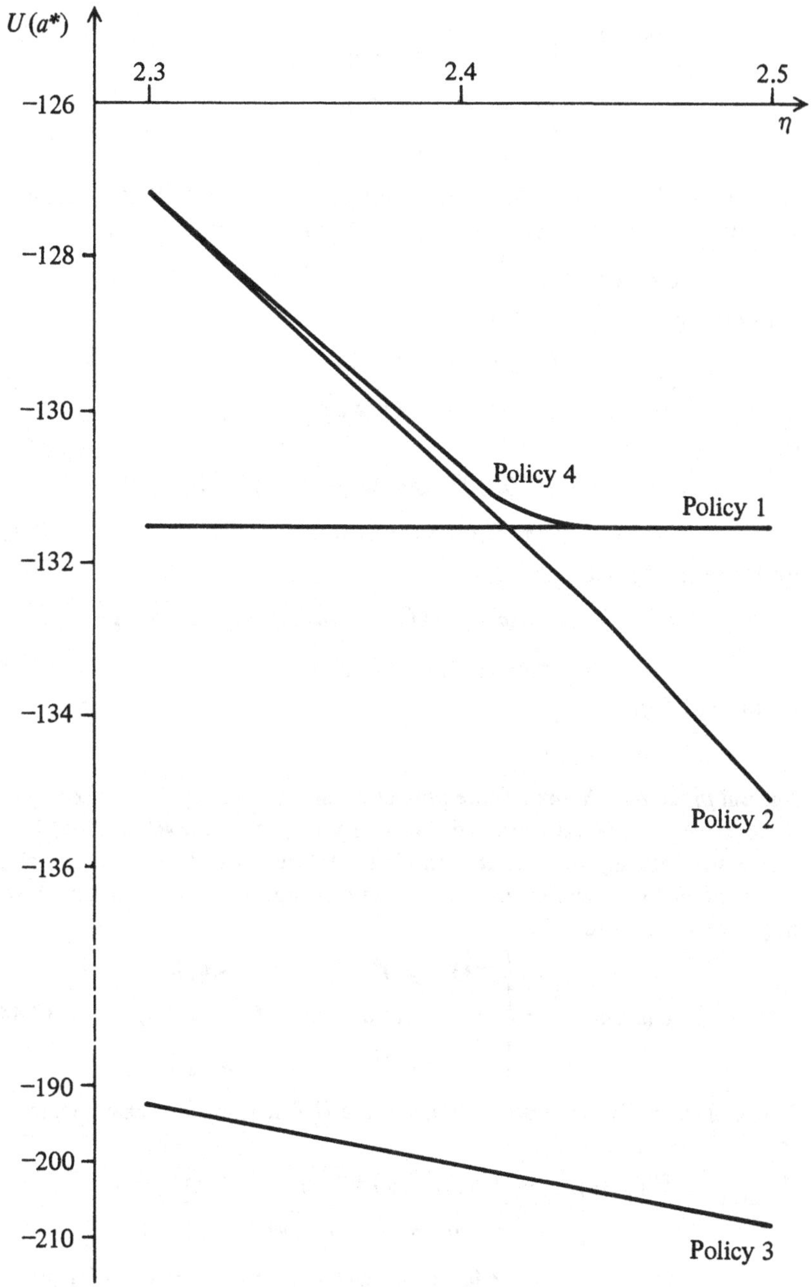

Fig. 3.5 Graphs of maximised expected utilities against η for the four policies.

$$U_j(a,y) = \begin{cases} y^j & (y < a), \\ 0 & (y \geqslant a); \end{cases} \tag{3.39}$$

$$U_j(a,y) = \begin{cases} (y-r)^j & (y < a), \\ 0 & (y \geqslant a). \end{cases} \tag{3.40}$$

The first is a form best suited to the investigation of discrete Y, the second and third are the better forms for the investigation of continuous Y.

Example 3.1

Linear utility point prediction. For

$$U(a,y) = \begin{cases} -\xi(a-y) & (y < a), \\ -\eta(y-a) & (y \geqslant a), \end{cases} \tag{3.41}$$

we have

$$U(a,y) = \xi U_1(a,y) - \xi a U_0(a,y) + \eta a\{1 - U_0(a,y)\} \\ -\eta\{U_1(\infty,y) - U_1(a,y)\} \tag{3.42}$$

for forms (3.38) and (3.39); and

$$U(a,y) = \xi U_1(a,y) - \xi(a-r)\, U_0(a,y) + \eta(a-r)\{1 - U_0(a,y)\} \\ -\eta\{U_1(\infty,y) - U_1(a,y)\} \tag{3.43}$$

for form (3.40).

Example 3.2

Interval prediction. When a finite prediction interval (a_1, a_2) is required and when losses are quadratic in the distance of y from the interval, and also increase quadratically with distance inside the interval (the deeper y is contained in the interval the more we may have used too large a prediction interval) we may have a utility function

$$U(a_1, a_2, y) = \begin{cases} -(y-a_1)^2 & (y < a_1), \\ -(y-a_1)(a_2-y) & (a_1 \leqslant y \leqslant a_2), \\ -(y-a_2)^2 & (y > a_2). \end{cases} \tag{3.44}$$

This is again easily expressed in terms of the U_j functions. For example, in terms of the form (3.39),

$$U(a_1, a_2, y) = -2U_2(a_1, y) + (3a_1 + a_2)U_1(a_1, y) \\ -a_1(a_1 + a_2)U_0(a_1, y) + 2U_2(a_2, y) \\ -(a_1 + 3a_2)U_1(a_2, y) + a_2(a_1 + a_2)U_0(a_2, y) \\ +2a_2 U_1(\infty, y) - a_2{}^2 - U_2(\infty, y). \tag{3.45}$$

Table 3.3 *Evaluation of* $U_j(a)$ *for standard predictive distributions*

Predictive distribution	$U_j(a, y)$	$U_j(a)$
BeBi (N, G, H)	(3.38)	$\frac{N!}{(N-j)!}\frac{B(G+j, H)}{B(G, H)} P_{hy}(N+G+H-1, G+a-1, N-j, a-1-j)$
NeBi $\left(G, \frac{K}{K+H}\right)$	(3.38)	$\frac{(G+j-1)!}{(G-1)!}\left(\frac{K}{H}\right)^j I_{H/(K+H)}(G+j, a-j)$
InBe (K, G, H)	(3.39)	$\frac{H^j B(K+j, G-j)}{B(K, G)} I_{a/(H+a)}(K+j, G-j)$
St $\left\{G, B, \left(\frac{1}{K}+\frac{1}{C}\right)\frac{H}{G}\right\}$	(3.40) $r = B$	$\begin{cases} \frac{1}{2}(-1)^j\left\{\left(\frac{1}{K}+\frac{1}{C}\right)H\right\}^{j/2} \frac{B\{\frac{1}{2}(j+1), \frac{1}{2}(G-j)\}}{B(\frac{1}{2}, \frac{1}{2}G)}[1-I_\gamma\{\frac{1}{2}(j+1), \frac{1}{2}(G-j)\}] & \text{if } a \leqslant B \\ \frac{1}{2}(-1)^j\left\{\left(\frac{1}{K}+\frac{1}{C}\right)H\right\}^{j/2} \frac{B\{\frac{1}{2}(j+1), \frac{1}{2}(G-j)\}}{B(\frac{1}{2}, \frac{1}{2}G)}[1+(-1)^j I_\gamma\{\frac{1}{2}(j+1), \frac{1}{2}(G-j)\}] & \text{if } a > B \end{cases}$ where $\gamma = \frac{(a-B)^2}{\left(\frac{1}{K}+\frac{1}{C}\right)H + (a-B)^2}$
$\frac{CKG}{(C+K)H}\left(1 + \frac{\omega_y(B, C, K)}{H}(y-B)\right)^{-(G+1)}$	(3.40) $r = B$	$\begin{cases} (-1)^j \frac{GK}{C+K}\left(\frac{H}{C}\right)^j B(j+1, G-j)\left\{1 - I_{(B-a)/[B-a+(H/C)]}(j+1, G-j)\right\} & \text{if } a \leqslant B \\ (-1)^j \frac{GK}{C+K}\left(\frac{H}{C}\right)^j B(j+1, G-j)\left\{1 + (-1)^j\left(\frac{C}{K}\right)^{j+1} I_{(a-B)/[a-B+(H/K)]}(j+1, G-j)\right\} & \text{if } a > B \end{cases}$

Integrating $U(a, y)$ or $U(a_1, a_2, y)$ with respect to $p(y|x)$ over Y to obtain the expected utilities $U(a)$ and $U(a_1, a_2)$ is a linear operation, and so we get precisely the forms (3.42), (3.43) and (3.45) above with the y omitted for these expected utilities in terms of

$$U_j(a) = \int_Y U_j(a, y) p(y|x) \mathrm{d}y. \tag{3.46}$$

It is therefore useful to have a catalogue of $U_j(a)$ corresponding to the simple $U_j(a, y)$ utility functions and the standard predictive distributions. These are provided in table 3.3.

3.6 An alternative formulation

The expected utility $U(a)$, which is the criterion for optimum prediction, can be expressed in an alternative form. For, under regularity conditions allowing a change of order of integration, we have the following development:

$$\begin{aligned} U(a) &= \int_Y U(a, y) p(y|x) \mathrm{d}y \\ &= \int_Y U(a, y) \left\{ \int_\Theta p(y|\theta) p(\theta|x) \mathrm{d}\theta \right\} \mathrm{d}y \quad \text{by (2.4)} \\ &= \int_\Theta \left\{ \int_Y U(a, y) p(y|\theta) \mathrm{d}y \right\} p(\theta|x) \mathrm{d}\theta \\ &= \int_\Theta V(a, \theta) p(\theta|x) \mathrm{d}\theta, \end{aligned} \tag{3.47}$$

where

$$V(a, \theta) = \int_Y U(a, y) p(y|\theta) \mathrm{d}y. \tag{3.48}$$

This formulation shows that we have a second way of viewing the problem because $U(a)$ has been expressed as the expectation of the induced utility $V(a, \theta)$ with respect to the posterior plausibility function $p(\theta|x)$. From any basic utility function $U(a, y)$, showing the assessment of utilities in terms of a and y, we can derive by (3.48) an equivalent induced utility $V(a, \theta)$, an assessment of utilities in terms of a and the index θ.

Not every function on $A \times \Theta$ necessarily arises from a function defined on $A \times Y$ by the relationship (3.48). For example

$$V(a_1, a_2, \theta) = \begin{cases} 1 - \gamma(a_2 - a_1) & \text{if } \int_{a_1}^{a_2} p(y|\theta) \mathrm{d}y \geqslant c, \\ -\gamma(a_2 - a_1) & \text{otherwise}, \end{cases} \tag{3.49}$$

is such a V function. This prompts us to ask the question: Are there any situations of prediction where a V formulation of utility applies and where there is no corresponding U specification? To answer this in the affirmative we need only envisage the future experiment f being repeated a large number of times and an interval (a_1, a_2) being regarded as of any use at all if it contains at least a proportion c of the outcomes from these replicates. We could then score 1 for success, 0 for failure. If the cost of an interval is proportional to its length we would then have the V function as specified in (3.49).

We shall see in §7.3 that the U and V specifications also provide a means of distinguishing between two fundamental types of informative prediction approaches.

History

The foundations of decision theory were laid down by Wald (1950) and the subject has expanded rapidly since then along both classical and Bayesian lines; see, for example, De Groot (1970), Ferguson (1967), Raiffa and Schlaifer (1961). The computational aids for more general utility functions in §3.5 and the alternative formulation of §3.6 are both discussed in Aitchison and Sculthorpe (1965).

Problems

3.1 For the situations of problems 2.1–2.4 find the all-or-nothing simple predictors, the linear loss simple predictors and the quadratic loss simple predictors as described in §3.1.

3.2 Complete the analysis of problem 1.2.

3.3 The research and development department associated with the extrusion process of problem 2.5 has been attempting to formulate the problem of predicting the number of flaws in 60 metre lengths of fibre as a decision problem. Two suggestions concerning the possible losses involved have been put forward:
(i) a prediction is useless unless it is correct,
(ii) the loss involved in a wrong prediction is equal to $10d^2$, where d is the difference between the predicted and the actual number of flaws.

What predictions would you advocate on the basis of these suggestions? What is the expected loss per prediction in case (ii)?

3.4 A delicatessen store owner has to sign a long-term contract whereby a fixed amount of a perishable new delicacy will be delivered to him daily. Any quantity sold on the day of delivery brings a profit of 3p per g whereas any quantity unsold involves a loss of 4p per g. After discussion with the store owner you feel that it is safe to assume that daily demand will be normally distributed but you remain vague about the values of the mean and variance parameters. During the 10 days before the contract has to be signed the daily amounts (g) demanded are

815, 920, 880, 830, 1125, 845, 990, 1200, 844, 1015.

Determine the optimum fixed daily order for the owner and the expected profit with this order.

3.5 In order to obtain favourable terms a theatre ticket agency has, at the end of the first week of a new show, to give a firm commitment to take a fixed number of theatre seats for each daily performance. The agency reckons that the daily demand of its clientele is Poisson-distributed but it is very vague about the mean parameter. The numbers for the first six performances are 8, 6, 3, 7, 2, 5. For each ticket sold the agency makes a profit of 50p, for each ticket unsold, a loss of £1. What fixed number of seats per day should the agency order, and what is its expected profit with this number?

3.6 Analyse the problem of interval prediction with the piecewise-linear utility function of the form (3.28) for the situations in problems 2.1 and 2.3.

3.7 Consider again the machine tool replacement problem (example 1.2, §§3.2, 3.4) and derive the optimal actions for the following policies (a) and (b).
(a) Send in the inspector at time a for a given interval of time of length T. Replace the machine tool as soon as it wears out or at the end of the interval whichever is earlier. If the machine tool has worn out before the inspector arrives the inspector replaces it and can be reallocated to another job at zero cost. Otherwise he is assigned for the whole period.
(b) As for policy (a) except that the inspector is contracted from outside and cannot be reallocated to another job.

Re-examine the relevant policies if a machine tool which is replaced while it is still functioning can be sold and some of its cost recouped, i.e. $\zeta < 0$.

3.8 In a certain country income tax policy is to tax only incomes above k, the tax being a fixed amount q. There is now a proposal to increase this tax from q to r for taxable incomes above l, where $l > k$, but to decrease the amount from q to s for taxable incomes below l. It is assumed that currently taxable incomes follow a $\text{Pa}(k, \theta)$ distribution, where little is known about the indexing parameter θ. A random sample of n currently taxable incomes shows recorded incomes $x_1, \ldots, x_n$.

Show that the proposal will provide a higher tax yield than current policy if $l < k(\bar{x}_g/k)^h$, where $h = n[\{(r-s)/(q-s)\}^{1/n} - 1]$ and $\bar{x}_g$ is the geometric mean of $x_1, \ldots, x_n$.

3.9 A production run of an industrial chemical process produces a large number of drums of sterilising liquid, which vary in acidity according to a $\mathrm{No}(\mu, \tau)$ distribution, where τ is known. The mean acidity μ of a run is not known, but μ is known to vary from run to run according to a $\mathrm{No}(b, c)$ distribution, where b and c are known. If an alkalising agent of strength a is injected into a drum of acidity y the resulting liquid has acidity $y - a$ if $a \leqslant y$ or alkalinity $a - y$ if $a > y$. The liquid is fully effective only when it is neutral; the loss of effectiveness from acid or alkaline liquid is proportional to the acidity or alkalinity, the factor of proportionality for acidity being thrice that for alkalinity. Administrative considerations dictate that each drum of a given run should receive the same strength of alkalising agent, and that only n drums from a run can be tested for acidity prior to treatment of the run.

Determine the optimum strength of alkalising agent for a run whose sampled drums have shown acidities $x_1, \ldots, x_n$.

3.10 Suppose that in a decisive prediction problem

$$A = Y = \mathbf{R}^n,$$

$$U(\mathbf{a}, \mathbf{y}) = -(\mathbf{a} - \mathbf{y})'\mathbf{M}(\mathbf{a} - \mathbf{y}),$$

where $\mathbf{M}$ is a positive definite matrix. Show that the optimum prediction is

$$\mathbf{a}^* = \mathrm{E}(\mathbf{y} \mid \mathbf{x})$$

and that the maximised expected utility is

$$-\operatorname{trace} \mathbf{M}\mathrm{V}(\mathbf{y} \mid \mathbf{x}),$$

where $\mathrm{E}(\mathbf{y} \mid \mathbf{x})$ and $\mathrm{V}(\mathbf{y} \mid \mathbf{x})$ are the mean and covariance matrix of the predictive distribution.

4
Informative prediction

4.1 Introduction

As already pointed out in chapter 2 the essence of prediction analysis from the Bayesian point of view is the construction of the predictive density function $p(y|x)$. When the appropriateness of predictions can be assessed in terms of a utility function the decisive prediction approach of chapter 3 resolves the prediction problem by maximising expected utility, expectation being taken with respect to the predictive density function. When there is difficulty in specifying the utility function and yet we want to convey some summary form of information about the plausible outcomes of the future experiment f, some principle other than the maximisation of expected utility must be introduced. To draw a clear distinction we suggest the term informative prediction to describe the use of any principle of prediction which does not require the specification of a utility function. As in previous chapters we use X and Y to denote the sample spaces of the informative experiment e and the future experiment f, and $\mathcal{Y}$ to denote the event space of f or the class of measurable subsets of Y.

Definition 4.1

Informative predictor. An *informative predictor* is a function

$$\delta : X \to \mathcal{Y} \tag{4.1}$$

which satisfies some probabilistic relation based on the probability measures associated with $p(x|\theta)$, $p(y|\theta)$ and, in the case of a Bayesian informative predictor, $p(\theta)$.

An informative predictor thus instructs the experimenter what region of the future experiment sample space he must use if his informative experiment yields x: he must use the informative prediction $\delta(x)$.

The object of an informative prediction region is clearly to narrow the range of possibility of a future observation from Y, so that receivers of this information may more readily plan or take appropriate action. Thus a manufacturer of components may wish to give to his customers some indication of the likely range of values within which a component characteristic (such as electrical resistance, lifetime) may lie. Again, a medical research worker who has devised a method of determining the urinary excretion rate of some steroid metabolite will wish to convey to other workers the 'normal range': he wants to provide a prediction interval within which other workers can be reasonably sure that most excretion rates of healthy persons lie. There is in such predictions, as in all statistical problems, a conflict between usefulness and validity. If we want to be absolutely sure that our prediction region will capture the actual outcome of a future experiment then we should quote Y as our prediction region. If we want to convey an extremely useful prediction then we should attempt to say that some specific $a \in Y$ will happen at the future performance, and probably place a forlorn hope on this point prediction a actually occurring. Clearly some compromise between these two extremes is necessary. In chapter 3 we saw how this can be achieved, when the effects of the conflict may be expressed by a suitable utility function, through the sophisticated yet simple principles of statistical decision theory. In the present and the following two chapters we study more primitive, and in some respects more complicated, analyses of such situations. These analyses have arisen to meet the needs of situations where it is difficult to set down a specific utility function – for example, where a manufacturer is trying to meet the needs of many customers.

In the present chapter we first discuss various aspects of such informative prediction from a Bayesian viewpoint using the predictive distribution. We then set the basis of the non-Bayesian or frequentist approach to informative prediction by introducing the general concept of the coverage distribution. Study of this distribution leads to two main types of prediction which have come to be known under the general heading of statistical tolerance regions. The two types – mean coverage and guaranteed coverage tolerance predictions – are separately developed in chapters 5 and 6.

4.2 Bayesian informative prediction

From a Bayesian viewpoint the natural way to assess the effectiveness of an informative prediction a is in terms of the probability that a future outcome lies in a. This probability, assessed in terms of the predictive distribution, we term the Bayesian cover provided by a.

Definition 4.2

Bayesian cover. The informative prediction $a \subset Y$ is said to have *Bayesian cover* κ if and only if

$$P(a|x) \equiv \int_a p(y|x)\mathrm{d}y = \kappa. \qquad (4.2)$$

Example 4.1

Bayesian cover of replacement policy. The Bayesian cover provided by the optimum prediction $(a_1{}^*, a_2{}^*)$ associated with policy 4 of §3.4 is, when condition (3.35) holds,

$$\int_{a_1^*}^{a_2^*} p(y|x)\mathrm{d}y = \left\{\frac{H}{H+a_1{}^*}\right\}^G - \left\{\frac{H}{H+a_2{}^*}\right\}^G$$

$$= \frac{\xi}{\xi+\eta} - \left\{\frac{(\eta-\xi)H}{\zeta G}\right\}^G$$

$$= 0.35 \tag{4.3}$$

for $\xi = 1.8$, $\eta = 2.4$, $\zeta = 54$. The practical interpretation of Bayesian cover here is the assessed probability that a machine tool will actually wear out during the attendance period. We are prepared to accept a low cover or probability 0.35 that the attendant will be present when the tool fails because of relatively high attendance costs.

There may be many predictions providing Bayesian cover equal to a specified value κ. A natural way of removing this ambiguity is to build up a prediction region by placing into it the most plausible values of y. This leads to the following formal definition.

Definition 4.3

Most plausible Bayesian prediction. A prediction a is a *most plausible Bayesian prediction* of cover κ if a has the form

$$a = \{y : p(y|x) \geqslant \gamma\}, \tag{4.4}$$

where γ is determined by

$$P\{a|x\} = \kappa. \tag{4.5}$$

We have already met an example of such Bayesian prediction in (3.27) for the all-or-nothing decisive interval prediction of (3.26). This correspondence gives us an interpretation for γ in (4.4), as the cost per unit interval in the equivalent all-or-nothing decisive prediction. Fig. 4.1 also gives a diagrammatic view of the determination of γ in the construction of a most plausible Bayesian prediction a of cover κ: we must adjust the horizontal line at height γ until the shaded area above the corresponding interval a and under the prediction density curve is κ.

Example 4.2

Most plausible Bayesian predictor for tool lifetime. Fig. 4.2 shows the graph of the predictive density function (3.12) associated with the machine tool

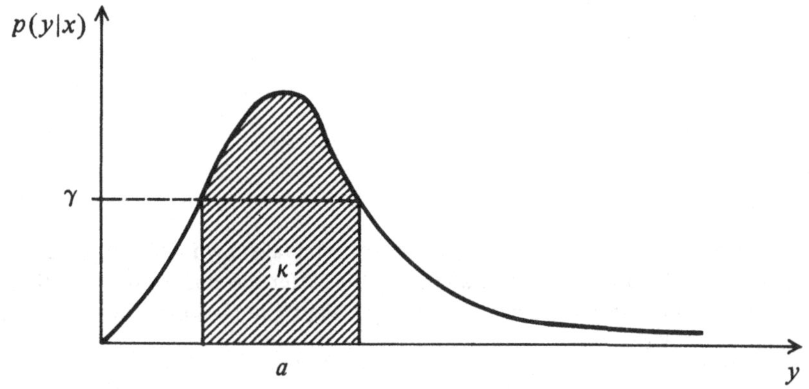

Fig. 4.1 A most plausible Bayesian prediction of cover κ.

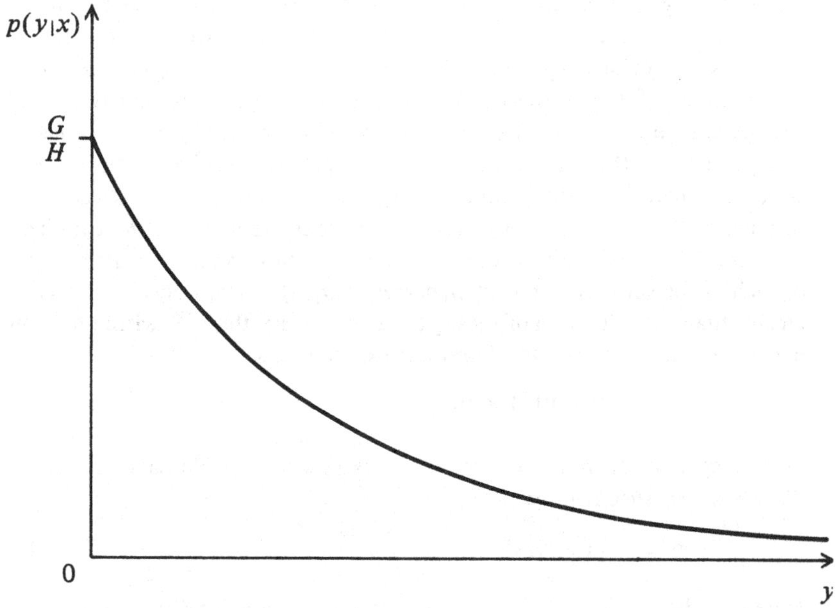

Fig. 4.2 Predictive density function (3.12) for the machine tool problem.

replacement problem of §§3.2 and 3.4. From the shape of the graph we see immediately that any interval a satisfying (4.4) must be of the form $(0, a)$ and a is determined by (4.5):

$$\int_0^a \frac{GH^G}{(H+y)^{G+1}}\,dy = \kappa,$$

so that

$$a = H\{(1-\kappa)^{-1/G} - 1\}. \tag{4.6}$$

Moreover

$$\gamma = G(1-\kappa)^{1+(1/G)}/H. \tag{4.7}$$

To obtain 95 per cent cover with this type of interval we would have to be prepared to attend the tool until 257.8 minutes after the start of its operation. The cost per minute associated with the equivalent all-or-nothing interpretation is then $\gamma = 0.0005$.

4.3 Region of previous experience

An interesting application of most plausible Bayesian prediction occurs in finding statistical expression for such terms as 'within previous experience' and 'outside previous experience'. Suppose that $\mathbf{x} = (x_1, \ldots, x_n)$ is a set of the outcomes of n replicates of some basic experiment, and we are assessing whether an outcome y from a newly performed experiment f can reasonably be regarded as within the experience of the n previous results. We first construct the predictive density function $p(y\,|\,\mathbf{x})$ which provides us with a measure of the plausibility of y based on previous experience $\mathbf{x}$. We can then compare this with the plausibility, as assessed by this predictive distribution, of each of the actually observed x_i, namely $p(x_i\,|\,\mathbf{x})$. If and only if $p(y\,|\,\mathbf{x})$ is greater than the minimum of these $p(x_i\,|\,\mathbf{x})$ do we say that y is within previous experience. Thus our *region of previous experience* is

$$a = \{y : p(y\,|\,\mathbf{x}) \geqslant \min_i p(x_i\,|\,\mathbf{x})\}. \tag{4.8}$$

Such a region of previous experience is thus a most plausible Bayesian prediction with γ specified by

$$\gamma = \min_i p(x_i\,|\,\mathbf{x}). \tag{4.9}$$

Note that the concept of the region of previous experience differs from that of range of previous experience defined as

$$(\min x_i, \max x_i)$$

for the one-dimensional case. For instance in example 4.2 the region of previous experience is (0, 290) whereas the range is (4, 290). Indeed the short lifetimes omitted from the range are the most plausible under the predictive distribution.

Cover of the region of previous experience. It is sometimes of interest to compute the cover associated with a prediction region. The interpretation of cover here is simply the probability that a new observation will fall within previous experience. For example we consider the multivariate normal situations in which we are led to predictive distributions of generalised Student form. Suppose that

$$p(\mathbf{y}|\mathbf{x}) \quad \text{is} \quad \mathrm{St}_d(k, \mathbf{b}, \mathbf{c}).$$

Then if we write

$$\lambda = \max_i (\mathbf{x}_i - \mathbf{b})'\mathbf{c}^{-1}(\mathbf{x}_i - \mathbf{b}) \tag{4.10}$$

the region of previous experience simplifies to

$$a = \{\mathbf{y} : (\mathbf{y} - \mathbf{b})'\mathbf{c}^{-1}(\mathbf{y} - \mathbf{b}) \leqslant \lambda\} \tag{4.11}$$

and so is the interior of an ellipsoid of concentration of the Student distribution.

The cover is given by

$$\int_a p(\mathbf{y}|\mathbf{x})d\mathbf{y}.$$

Although this integral appears complicated it can be reduced to a standard function by the following series of elementary transformations and steps.

(i) $\mathbf{y} - \mathbf{b} = \sqrt{(k)}\mathbf{W}\mathbf{z}$ where $\mathbf{W}'\mathbf{c}^{-1}\mathbf{W} = \mathbf{I}_d$ is the unit matrix of order d, the dimension of the multivariate distribution.

(ii) $\mathbf{z} \to$ generalised spherical polar coordinates $(r, \psi_1, \ldots, \psi_{d-1})$.

(iii) Since the integrand factorises to $f(r)g(\Psi)$ and the region of integration is determined only by a restriction on r, we can easily integrate out $\psi_1, \ldots, \psi_{d-1}$, obtaining

$$\frac{2}{B\{\frac{1}{2}d, \frac{1}{2}(k-d+1)\}} \int_0^{\sqrt{(\lambda/k)}} \frac{r^{d-1}}{(1+r^2)^{(k+1)/2}} dr.$$

(iv) $u = r^2/(1+r^2)$ or $r = \sqrt{\{u/(1-u)\}}$ then gives

$$\int_a p(\mathbf{y}|\mathbf{x})d\mathbf{y} = \int_0^{\lambda/(\lambda+k)} \frac{u^{(d/2)-1}(1-u)^{\{(k-d+1)/2\}-1}}{B\{\frac{1}{2}d, \frac{1}{2}(k-d+1)\}} du$$
$$= I_{\lambda/(\lambda+k)}\{\tfrac{1}{2}d, \tfrac{1}{2}(k-d+1)\} \tag{4.12}$$

an incomplete beta function, as defined in (A19) of appendix I.

4.4 An application to metabolite excretion rates

Table 4.1 shows the urinary excretion rates (mg/24 h) of two steroid

Table 4.1 *Urinary excretion rates (mg/24h) of cortisol and cortisone for 27 cases of Cushing's syndrome with adrenal hyperplasia*

Patient no.	Cortisol	Cortisone	Value of quadratic form	Patient no.	Cortisol	Cortisone	Value of quadratic form
1	0.41	0.38	1.66	15	0.48	0.36	0.85
2	0.16	0.18	2.97	16	0.80	0.39	2.42
3	0.26	0.15	0.79	17	0.40	0.24	0.12
4	0.34	0.33	1.51	18	0.22	0.10	3.46
5	1.12	0.60	4.71	19	0.24	0.24	1.44
6	0.15	0.14	2.36	20	0.56	0.42	1.48
7	0.20	0.16	0.96	21	0.40	0.16	2.50
8	0.26	0.18	0.27	22	0.88	0.48	2.94
9	0.56	0.32	0.79	23	0.44	0.26	0.24
10	0.26	0.20	0.30	24	0.24	0.16	0.53
11	0.16	0.13	1.90	25	0.27	0.19	0.19
12	0.56	0.33	0.79	26	0.18	0.18	1.92
13	0.33	0.08	11.49	27	0.60	0.35	1.03
14	0.26	0.22	0.53				

metabolites, cortisol and cortisone, determined by paperchromatography for 27 patients diagnosed as having Cushing's syndrome with adrenal hyperplasia. There are other forms of Cushing's syndrome and other steroid metabolites which are relevant to the differential diagnosis of the syndrome (§ 11.1) but we restrict attention here to the construction of a region of previous experience for cortisol and cortisone, since this problem is sufficient to illustrate the technique.

The first task is to construct the predictive distribution. To cope with the skewness of the data we recognise the approximate lognormality and work throughout with logarithms (to the base 10) of the excretion rates. The transformed data then arise from $n = 27$ replicates of a multinormal $\text{No}_d(\boldsymbol{\mu}, \boldsymbol{\tau})$ experiment with dimension $d = 2$. These data can then be summarised sufficiently in the two-dimensional vector mean $\mathbf{m}$ and the 2×2 matrix $\mathbf{v}$ of corrected sums of squares and cross-products. We are thus concerned with case 6 of table 2.3 with $\mathbf{m}$ and $\mathbf{v}$ independently distributed as $\text{No}_2(\boldsymbol{\mu}, n\boldsymbol{\tau})$ and $\text{Wi}_2(n-1, \boldsymbol{\tau})$. We are interested in the problem of predicting the two excretion rates for a new hyperplasia patient, and so in a $\text{No}_2(\boldsymbol{\mu}, \boldsymbol{\tau})$ future experiment f. Following through the construction process of table 2.3 using the vague prior for $(\boldsymbol{\mu}, \boldsymbol{\tau})$ we arrive at the predictive distribution

$$\text{St}_2\left\{n-1, \mathbf{m}, \left(1+\frac{1}{n}\right)\frac{\mathbf{v}}{n-1}\right\}. \tag{4.13}$$

For the given data

$$n-1 = 26, \quad \mathbf{m} = \begin{bmatrix} -0.4650 \\ -0.6403 \end{bmatrix},$$

$$\frac{n(n-1)}{n+1}\mathbf{v}^{-1} = \begin{bmatrix} 47.8802 & -42.1236 \\ -42.1236 & 58.1050 \end{bmatrix}.$$

The values of the quadratic forms in (4.10) are then easily computed and are also shown in table 4.1. The maximum of these is $\lambda = 11.49$ and so the region of previous hyperplasia experience is bounded by the ellipse

$$47.8802(y_1 + 0.4650)^2 + 58.1050(y_2 + 0.6403)^2$$

$$- 84.2472(y_1 + 0.4650)(y_2 + 0.6403) = 11.49.$$

Fig. 4.3 shows the positions of the 27 cases in the cortisol–cortisone plane and the bounding ellipse. The cover provided by this region of previous experience is, by (4.12),

$$I_{0.306}(1, 12.5) = 0.99,$$

so that we would expect only about one in a hundred of hyperplasia patients

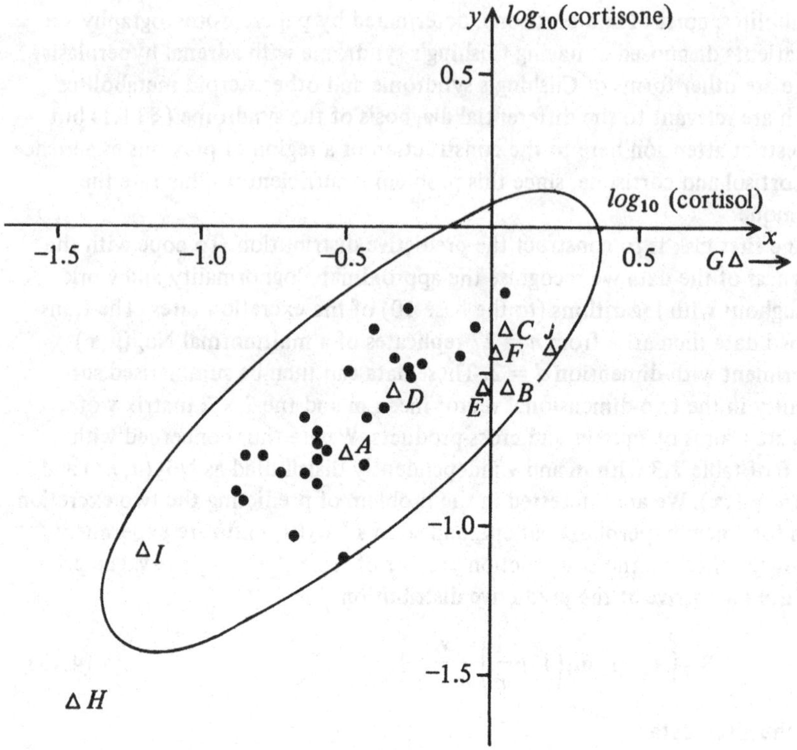

Fig. 4.3 Region of previous experience of 27 patients having Cushing's syndrome with adrenal hyperplasia and the relative positions of 10 other patients.

to have cortisol–cortisone values outside the ellipse. Table 4.2 shows the excretion rates for ten new patients, their quadratic form values and the confirmed diagnostic form of Cushing's syndrome for the eight patients found to be suffering from it. The point representations of the ten patients are shown in fig. 4.3. All six of the new hyperplasia cases A–F fall within the region of previous hyperplasia experience, and two non-hyperplasia cases G and H, fall outside the region. The fact that the other two non-hyperplasia cases fall within the hyperplasia region is a reminder that the construction of such a region of previous experience does not provide a statistical means of diagnosing between hyperplasia and non-hyperplasia. Any such differential diagnostic system must be built not only on hyperplasia experience but also on information about the excretion rates in non-hyperplasia cases. Regions of previous experience for the other forms of Cushing's syndrome do in fact intersect the hyperplasia region of fig. 4.3. What the construction of a region of hyperplasia

Table 4.2 *Urinary excretion rates (mg/24h) of cortisol and cortisone for ten new patients*

Patient	Cortisol	Cortisone	Value of quadratic form	Confirmed differential diagnosis
A	0.32	0.18	0.41	Hyperplasia
B	1.12	0.32	7.59	Hyperplasia
C	1.12	0.48	4.74	Hyperplasia
D	0.48	0.31	0.41	Hyperplasia
E	0.96	0.32	5.33	Hyperplasia
F	1.04	0.40	4.69	Hyperplasia
G	11.20	0.75	59.47	Ectopic carcinoma
H	0.04	0.03	17.57	Normal
I	0.07	0.10	9.40	Normal
J	1.60	0.40	11.19	Adrenal carcinoma

experience does provide is a means of monitoring new cases diagnosed as hyperplasia to ensure that they conform reasonably with previous experience of that particular form. See §11.4 for further discussion of monitoring for atypicality in diagnostic problems.

4.5 Bayesian coverage

If we apply a predictor δ to a number of informative experiments then the cover provided by the predictions supplied by δ will vary from informative experiment to informative experiment. For the purpose of studying this variability it is worth making a formal definition.

Definition 4.4

Bayesian coverage of a predictor. The *Bayesian coverage* of a predictor $\delta : X \rightarrow \mathcal{Y}$ is the statistic

$$\int_{\delta(x)} p(y \mid x) \mathrm{d}y \tag{4.14}$$

defined on the sample space X of the informative experiment. The *mean Bayesian coverage* of a predictor δ is then

$$\int_X p(x) \mathrm{d}x \int_{\delta(x)} p(y \mid x) \mathrm{d}y, \tag{4.15}$$

and can be interpreted as the long-run average Bayesian cover provided by the predictor if it is applied to a sequence of outcomes of the informative experiment.

Example 4.3

Machine tool replacement. If we use policy 3 in §3.2 we are expecting the machine to fail during the time-interval $(0, a^*)$ and so in effect are using an interval predictor

$$\delta(x) = \begin{cases} \text{interval prediction } \left\{0, \dfrac{\zeta G}{\eta - \xi} - (h + x)\right\} \text{ if } x < \dfrac{\zeta G}{\eta - \xi} - h, \\ \text{point prediction } \{0\} \text{ otherwise.} \end{cases}$$

Thus the Bayesian coverage of δ is

$$\begin{cases} 1 - \left\{\dfrac{(h + x)(\eta - \xi)}{\zeta G}\right\}^G & \text{if } x < \dfrac{\zeta G}{\eta - \xi} - h, \\ 0 & \text{otherwise.} \end{cases} \tag{4.16}$$

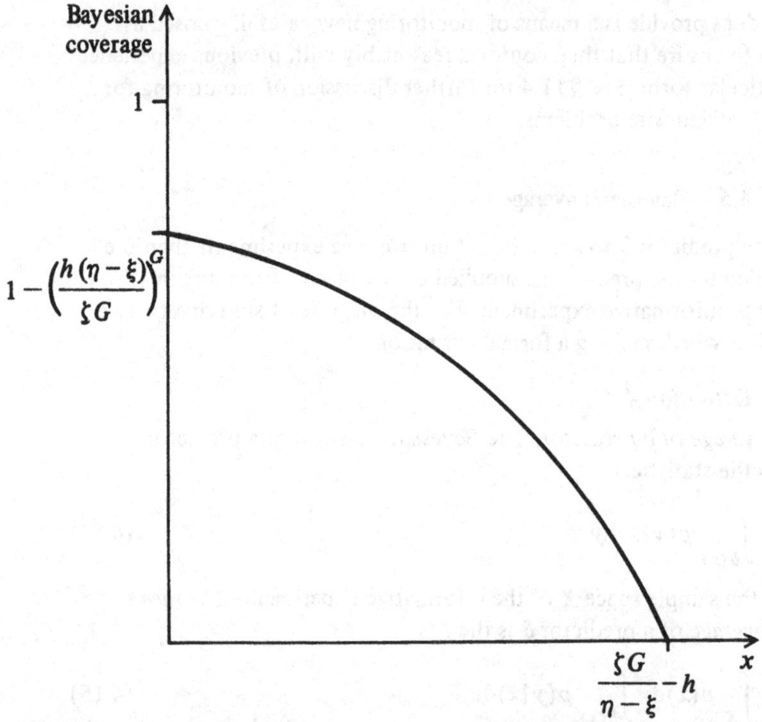

Fig. 4.4 Relationship of Bayesian coverage (4.16) to x for the machine tool problem.

The relationship of Bayesian coverage to x is shown in fig. 4.4. Here

$$p(x) = \text{InBe}(n, g, h)$$

and the mean coverage is

$$\int_0^{\frac{\zeta G}{\eta-\xi}-h} \left\{1-\left(\frac{\eta-\xi}{\zeta G}\right)^G (h+x)^G\right\} p(x)\mathrm{d}x$$

$$= I_{1-\rho}(n,g)-\frac{\rho^g(1-\rho)^n}{n\mathrm{B}(n,g)}, \tag{4.17}$$

where

$$\rho = \frac{h(\eta-\xi)}{\zeta(g+n)}. \tag{4.18}$$

4.6 Statistical tolerance regions: role of coverage distributions

Much research effort has been devoted to methods of constructing prediction regions under the restriction that no prior distribution on Θ is available. The resulting tolerance regions are made to satisfy certain probabilistic statements in much the same way that confidence intervals for an unknown parameter value are constructed. Unfortunately the more complicated predictive situation requires a more sophisticated probabilistic statement and this extra complexity is often a deterrent to potential users. We shall therefore attempt in this section to give a relatively fresh account of the principles in terms of the unifying concept of the coverage distribution. Also an adequate notation is required to make clear the meaning of the probabilistic statements that form the basis of statistical tolerance regions. We continue to label the informative experiment by e and the future experiment by f but now place additional identifiers on the informative and future experiment density functions $p_e(x|\theta)$ and $p_f(y|\theta)$ and the corresponding probability measures $P_e(\cdot|\theta)$ and $P_f(\cdot|\theta)$ over X and Y. We shall also require the product density function

$$p_{ef}(x,y|\theta) = p_e(x|\theta)p_f(y|\theta) \tag{4.19}$$

associated with the combined experiment (e,f) and the corresponding probability measure $P_{ef}(\cdot|\theta)$ over $X \times Y$. To emphasise that we are not now dealing with predictions or prediction regions based on predictive distributions we shall use the established term *tolerance region* or *tolerance prediction* in what follows.

Cover of a tolerance region. A basic concept in all work on tolerance regions is that of the cover provided by such a region $a \subset Y$. We define this to be $P_f(a|\theta)$, where $P_f(\cdot|\theta)$ denotes the probability measure associated with the future experiment f for which a prediction region is required. The relative frequency interpretation of cover is the following: if we use a as a tolerance region for a long run of independent repetitions of the future experiment then a will successfully cover or contain a proportion $P_f(a|\theta)$ of the outcomes from these repetitions. Note that the cover depends not only on the region a but also on the index θ. We take account of this in our formal definition.

Definition 4.5

Cover of a tolerance region. The *cover* of a tolerance region a with respect to θ (or, for the sake of brevity, at θ) is

$$P_f(a\,|\,\theta). \tag{4.20}$$

Example 4.4

Exponential case. Suppose that

$$p_e(x\,|\,\theta) = \theta \exp(-\theta x)(x>0), \quad p_f(y\,|\,\theta) = \theta \exp(-\theta y)(y>0). \tag{4.21}$$

Consider the tolerance interval $\delta(x) = (q_1 x, q_2 x)$, where $0 \leqslant q_1 \leqslant q_2$. Then the cover provided by this tolerance interval is

$$P_f\{(q_1 x, q_2 x)\,|\,\theta\} = \int_{q_1 x}^{q_2 x} \theta \exp(-\theta y)\,dy$$
$$= \exp(-\theta q_1 x) - \exp(-\theta q_2 x). \tag{4.22}$$

Coverage and coverage distribution of a tolerance predictor. In the preceding example a repetition of the informative experiment would almost certainly result in a different outcome x, and consequently a different tolerance interval $\delta(x)$ with different cover $P_f\{\delta(x)\,|\,\theta\}$. This inherent variability in x and the variability it induces in $P_f\{\delta(x)\,|\,\theta\}$ play a central role in the definition and construction of statistical tolerance regions. As a crucial first step towards the construction of such regions we therefore introduce the following definition.

Definition 4.6

Coverage and coverage distribution. For a predictor δ the *coverage* at θ is the random variable

$$P_f\{\delta(x)\,|\,\theta\} \tag{4.23}$$

and its distribution is termed the *coverage distribution*.

A predictor has a coverage distribution corresponding to each index θ. Since we do not know the true value of θ and are currently assuming that no prior distribution on Θ is available all we can hope to provide is a predictor δ whose coverage distributions all satisfy some desirable property. Since any coverage distribution is a distribution over the interval $[0, 1]$ we would ideally want all the coverage distributions to be concentrated on 1 (fig. 4.5a). Since this is clearly unattainable what we seek is some requirement on the coverage distributions which forces them to approximate to this ideal. These requirements are of two main types.

(1) Ensure that, for every $\theta \in \Theta$, the mean of the coverage distribution is

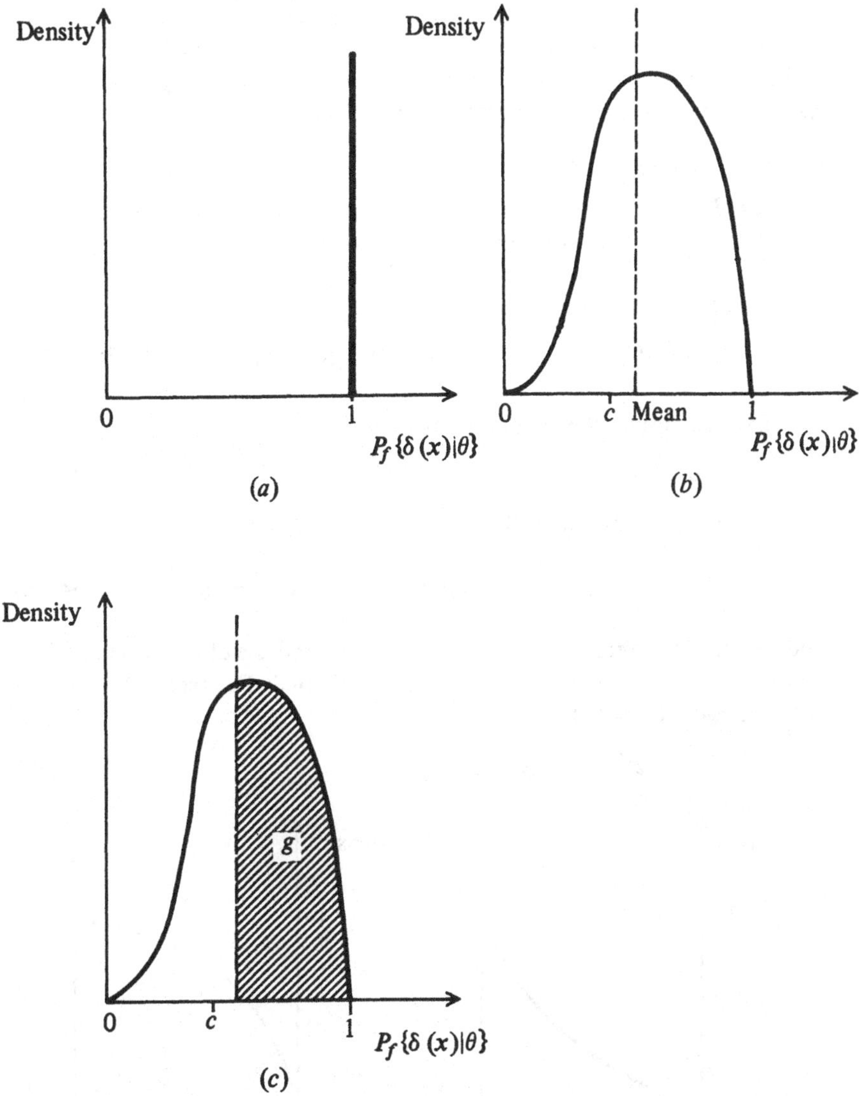

Fig. 4.5 Coverage distributions: (*a*) ideal, (*b*) mean coverage, (*c*) guaranteed coverage

reasonably high, say at least c (fig. 4.5b). This criterion produces what is commonly termed a *mean coverage tolerance region.*

(2) Ensure that, for every $\theta \in \Theta$, the bulk of the coverage distribution is above some specified cover value, c say. To guarantee that at least a proportion g of the coverage distribution lies above c for every possible θ is

essentially to place a restriction on a quantile rather than the mean. We require (fig. 4.5c) that the $(1-g)$-quantile lies at or to the right of c for every θ. Such a requirement leads to what may be termed a *quantile-guaranteed* or, more briefly, a *guaranteed coverage tolerance* region.

Example 4.5

Two simple coverage distributions. To illustrate the nature of coverage distributions we study two extremely simple cases.

(i) Exponential distribution. Suppose that

$$p_e(x|\theta) = \theta \exp(-\theta x)(x > 0), \quad p_f(y|\theta) = \theta \exp(-\theta y)(y > 0), \tag{4.24}$$

and that we consider the predictor δ defined by

$$\delta(x) = (qx, \infty), \tag{4.25}$$

where $q > 0$. The coverage is then

$$P_f\{\delta(x)|\theta\} = \int_{qx}^{\infty} \theta \exp(-\theta y)dy = \exp(-\theta qx), \tag{4.26}$$

and to find the coverage distribution we have to find the distribution of $z = \exp(-\theta qx)$, where x is exponentially distributed. The density function of the coverage is therefore

$$(1/q)z^{(1/q)-1} \quad (0 < z < 1). \tag{4.27}$$

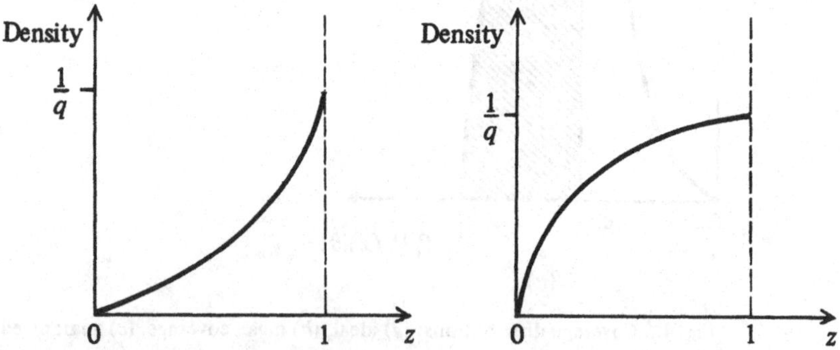

(*a*) (*b*)

Fig. 4.6 Graph of density function (4.27) for (*a*) $0 < q < \frac{1}{2}$ (*b*) $q \geq \frac{1}{2}$.

Fig. 4.6 shows the graph of this density function. Since

$$E(z) = 1/(q+1)$$

we can attain the mean coverage requirement of (1) by setting

$$q = (1-c)/c. \tag{4.28}$$

Also since

$$\int_c^1 \frac{1}{q} z^{(1/q)-1}\, dz = 1 - c^{1/q}$$

we obtain the quantile property of (2) by setting

$$q = \frac{\log c}{\log(1-g)}. \tag{4.29}$$

For this predictor the coverage distribution does not depend on θ. This will not always be the case. For instance if we consider the predictor δ defined by

$$\delta(x) = (q + x, \infty), \tag{4.30}$$

the density function of the coverage is given by

$$\begin{cases} \exp(\theta q) & (0 < z \leqslant \exp(-\theta q)), \\ 0 & (\exp(-\theta q) < z < 1), \end{cases} \tag{4.31}$$

which depends crucially on θ.

We have thus within this simple example all the ingredients which are present in the more complex situations dealt with later in chapters 5 and 6.

(ii) Binomial trial. A second exceedingly simple illustrative example is specified by the informative and future experiment density functions:

$$\begin{aligned} p_e(0|\theta) &= \theta, \quad p_e(1|\theta) = 1-\theta, \\ p_f(0|\theta) &= \theta, \quad p_f(1|\theta) = 1-\theta; \end{aligned} \tag{4.32}$$

that is, both experiments are simple binomial trials. Let us consider the predictor δ specified by

$$\delta(0) = \{0\}, \quad \delta(1) = \{1\},$$

and expressing the view that what has happened in the informative experiment will happen in the future experiment. The coverage statistic then takes two possible values

$$\begin{aligned} P_f\{\delta(0)|\theta\} &= p_f(0|\theta) = \theta, \\ P_f\{\delta(1)|\theta\} &= p_f(1|\theta) = 1-\theta, \end{aligned} \tag{4.33}$$

and since these values are taken with (informative experiment) probabilities θ and $1-\theta$, we see that the coverage z has density function $p(z)$ specified by

$$p(\theta) = \theta, \quad p(1-\theta) = 1-\theta. \tag{4.34}$$

History

The concept of Bayesian cover is used by Aitchison and Sculthorpe (1965). Regions of previous experience for the multinormal case are simply ellipsoids of concentration, and so have a long ancestry. For their application to a medical situation see, for example, Ferriss *et al* (1970). The assessment of the probability (4.12) of obtaining a new case outside previous experience through the use of the predictive distribution is reported by Aitchison and Kay (1975).

Problems

4.1 Given that

$p(\theta)$ is $\mathrm{Ga}(g, h)$,

$p(x|\theta)$ is $\mathrm{Ga}(n, \theta)$,

$p(y|\theta)$ is $\mathrm{Ex}(\theta)$,

find $p(y|x)$, $\mathrm{E}(y|x)$ and $\mathrm{V}(y|x)$.

What are the Bayesian covers of the following prediction intervals?

(i) The optimum prediction interval of the form $(0, a)$ with respect to the utility function

$$U(a,y) = \begin{cases} -\xi(a-y) & (y < a), \\ -\eta(y-a) & (y \geqslant a). \end{cases}$$

(ii) The prediction interval $\{k(h+x), (h+x)/k\}$, where $k < 1$. Show that, if $n > 2$, the Bayesian cover lies between $(1-k)/(1+k)^{g+n-1}$ and $1/(1+k)^{g+n}$.

(iii) The prediction interval $\{0, \mathrm{E}(y|x)\}$. Show that as $n \to \infty$ the Bayesian cover tends to $1 - e^{-1}$. Can you suggest an alternative explanation of this simple result?

4.2 Complete problem 1.1 interpreting 'range of normality' as
(i) the most plausible Bayesian prediction interval of cover 0.95,
(ii) the region of previous experience,
in each case assuming that the biparietal diameters are normally distributed and making use of an appropriate vague prior distribution.

4.3 For the predictive distributions associated with problems 2.1–2.4 obtain most plausible Bayesian prediction intervals of cover κ.

4.4 For problem 2.9 find most plausible Bayesian prediction intervals of cover 0.99 for

(i) the mean of the 15 further bearings,
(ii) the standard deviation of the 15 further bearings.

Obtain the interval of diameter experience associated with the sample of 10 ball bearings. What is the probability that the diameter of another ball bearing from the batch will be outside previous experience?

4.5 For problem 2.10 obtain a most plausible Bayesian prediction of cover 0.90 for the lifetime of the eleventh component.

4.6 For problem 2.12 obtain a most plausible Bayesian prediction of cover 0.95 for the total yield of the eight tomato plants all grown in compost with additive at strength 5.5.

4.7 The times (hours) to first breakdown of 20 machines were measured in an informative experiment and were as follows.

4	10	62	119	74	24	13	29	19	18
57	23	47	409	19	208	13	209	46	188

We assume that these are independent observations on $\text{Ex}(\theta)$ random variables where little prior information on θ is available. Five similar machines are to be used simultaneously (and independently) in a laboratory and we wish to give prediction intervals for

(i) the time to the first failure, $y_{[1]}$,
(ii) the time to the last failure, $y_{[5]}$,
(iii) the total running time of the machines, $\sum_1^5 y_i$.

Derive the predictive density functions in each case and hence obtain Bayesian predictive intervals with cover 0.95.

4.8 Find the two-dimensional regions of previous experience for
(a) the adenoma patients,
(b) the bilateral hyperplasia patients,
associated with the K and CO_2 measurements *only* of table 1.6.

What are the probabilities
(i) that a new adenoma patient will fall within previous adenoma experience;
(ii) that a new bilateral hyperplasia patient will fall within previous bilateral hyperplasia experience?

How many bilateral hyperplasia patients fall within previous adenoma experience, and how many adenoma patients fall within previous bilateral hyperplasia experience?

4.9 The urinary excretion rates (mg/24h) of three steroid metabolites A, B and C are shown in the following table for 12 normal healthy adults. Construct a region of previous experience for this set of data. Assess the probability that another normal healthy adult will have metabolite results outside this region of previous experience.

A	*B*	*C*
3.0	4.2	1.4
5.7	8.0	3.2
1.6	7.7	3.4
4.0	4.7	2.4
4.3	5.5	4.5
3.3	3.5	6.8
6.1	2.8	4.8
3.2	2.8	3.4
3.5	5.7	4.0
2.2	8.7	5.3
4.9	9.4	3.7
2.6	5.5	4.1

For the following three new cases first attempt to assign an index of abnormality intuitively, and then compute such an index.

A	*B*	*C*
6.0	5.1	7.0
4.0	8.2	3.8
5.5	4.7	6.5

4.10 Let $\delta(x)$ denote a most plausible Bayesian prediction of cover κ based on data x from an informative experiment. What is the coverage distribution of the predictor δ and what is the Bayesian coverage?

4.11 Find the density function of the coverage of the predictor δ defined by

$$\delta(x) = (qx, \infty),$$

where $q > 0$, for the prediction problem:

$$p_e(x|\theta) = \text{Ga}(k, \theta) \quad p_f(y|\theta) = \text{Ex}(\theta).$$

Is it possible to obtain mean coverage and guaranteed coverage tolerance predictors for this case?

4.12 An attempt has been made to design 'self-destructive' components in such a way that each has a minimum lifetime μ and that, subsequent to that minimum period, each component follows a failure pattern which is a Poisson process of high intensity τ. For any production run control has not yet reached the stage where μ and τ can be predetermined, but study of a number of trial production runs suggests that the joint variability in μ and τ follows an

ElGa(b, c, g, h) distribution. What is now proposed is that from the large number of components of each production run a random sample of n components should be observed and the lifetimes $x_1, \ldots, x_n$ determined. Then for each production run it is proposed to quote an interval within which 99 per cent of lifetimes for that run will lie. It is understood that for any one production run this may be misleading but it is hoped that over a number of batches this policy of quoting intervals will be effective.

Advise the management on how to construct such intervals. How would your answer be affected if one and only one component (say the nth) was still functioning at time t when an interval had to be quoted?

For an entirely new trial production process about whose characteristics μ and τ little is known the lifetimes (days) of a random sample of 10 components from the large production run are given below. What proportion of the other components do you assess have lifetimes within the same range?

195, 160, 243, 212, 188, 160, 157, 173, 169, 162.

4.13 In a certain factory each production run produces a large number of items. For each of k such production runs some items have been tested and found to have the following characteristics (measurements of quality).

Production run	Measurements of quality
1	$w_{11}, w_{12}, \ldots, w_{1n_1}$
2	$w_{21}, w_{22}, \ldots, w_{2n_2}$
.	. .
.	. .
.	. .
k	$w_{k1}, w_{k2}, \ldots, w_{kn_k}$

The possibility that the distribution of the characteristic differs with production run has now been realised. It is therefore proposed to test n items from each production run and to make some statement about the distribution of characteristics associated with that production run. For a production run with test components giving characteristic values

$$x_1, x_2, \ldots, x_n$$

what can be said about the distribution of the characteristic in the large number of untested items from that run?

5
Mean coverage tolerance prediction

5.1 Introduction

The motivation underlying the search for mean coverage tolerance predictors is an attempt to ensure that the average cover provided by the predictor does not fall too low. Since the mean of the coverage distribution of a predictor depends in general on θ this aim can be secured only by constructing the predictor in such a way that none of these possible means fall below some preassigned cover, say c. We can express this idea formally in terms of the following definition.

Definition 5.1

Mean coverage tolerance predictor. A predictor δ is a *tolerance predictor of mean coverage c* if

$$\inf_{\Theta} E[P_f\{\delta(\cdot)|\theta\}] = c, \tag{5.1}$$

where

$$\mathrm{E}[P_f\{\delta(\cdot)|\theta\}] = \int_X P_f\{\delta(x)|\theta\}p_e(x|\theta)\,\mathrm{d}x. \tag{5.2}$$

For the case of discrete distributions it is necessary to relax the definition, the equality sign being replaced by $\geqslant$.

Also the concept of *similarity* may apply.

Definition 5.2

Similar mean coverage tolerance predictor. A predictor δ is a *tolerance predictor of similar mean coverage c* if

$$\mathrm{E}[P_f\{\delta(\cdot)|\theta\}] = c \text{ for every } \theta \in \Theta. \tag{5.3}$$

We have extended the usual definition of mean coverage tolerance predictors, the similar form of definition 5.2, to the more general form of definition 5.1. While for some standard situations such as the normal and gamma cases similar mean coverage tolerance predictors can be found there are other situations, for example the binomial case, where no such similar mean coverage tolerance predictors may exist. It is for this reason that we introduce the less demanding requirement (5.1) of definition 5.1. The distinction between the two definitions is analogous to the concepts of size and exact or similar size of critical regions in the classical theory of hypothesis testing.

Example 5.1

Two simple mean coverage tolerance predictors

(i) Exponential distribution. We have already seen in example 4.5 (i) a case of a similar mean coverage tolerance predictor. Indeed there are circumstances in which more than one such predictor may exist.

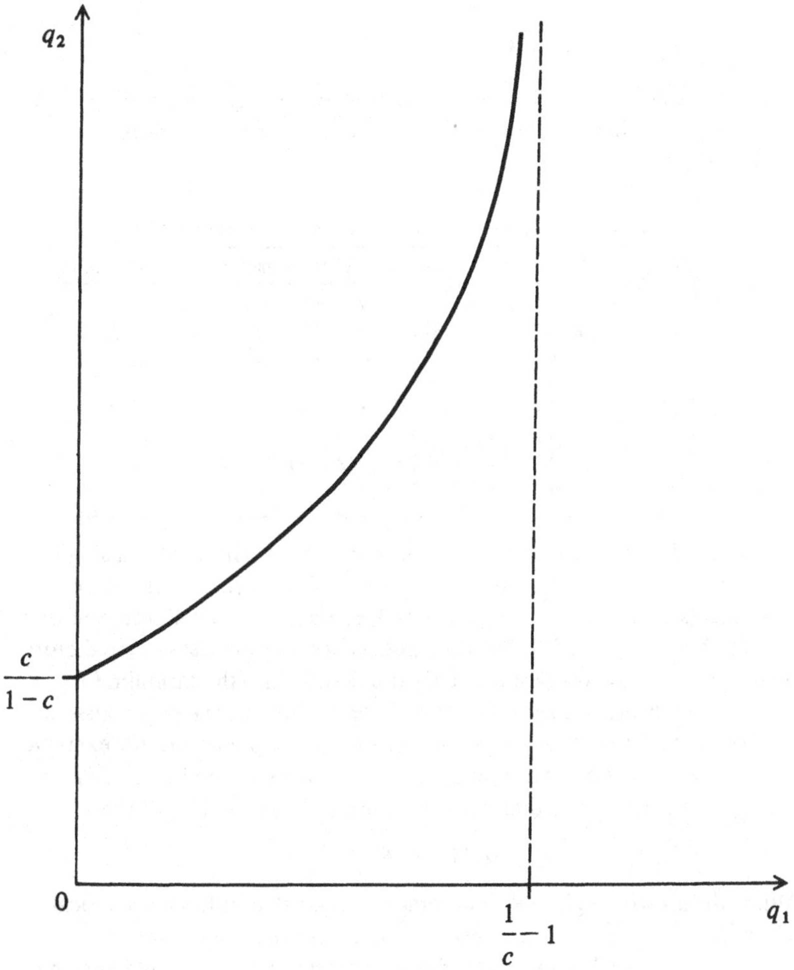

Fig. 5.1 Locus of (q_1, q_2) for which $\frac{1}{q_1+1} - \frac{1}{q_2+1} = c$ for given c.

Consider the predictor

$$\delta(x) = (q_1 x, q_2 x)$$

discussed in example 4.4. From our previous work we easily calculate the integral in (5.2) as

$$\int_0^\infty \{\exp(-\theta q_1 x) - \exp(-\theta q_2 x)\}\,\theta \exp(-\theta x)dx = \frac{1}{q_1+1} - \frac{1}{q_2+1}.$$

Since this is independent of θ, to satisfy (5.3) we need only select values of q_1 and q_2 such that $q_2 > q_1$ and

$$\frac{1}{q_1+1} - \frac{1}{q_2+1} = c. \tag{5.4}$$

It is easily seen from fig. 5.1 that this is possible for any value c in $0 < c < 1$ and indeed that there is an embarrassingly large number of such (q_1, q_2) combinations.

Table 5.1 *The eight possible predictors δ_i for the simple binomial example*

i	$\delta_i(0)$	$\delta_i(1)$	Mean coverage at θ	Infimum
1	$\{0\}$	$\{0, 1\}$	$\theta^2 + (1-\theta)$	$\frac{3}{4}$
2	$\{0, 1\}$	$\{1\}$	$\theta + (1-\theta)^2$	$\frac{3}{4}$
3	$\{0\}$	$\{1\}$	$\theta^2 + (1-\theta)^2$	$\frac{1}{2}$
4	$\{0\}$	$\{0\}$	θ	0
5	$\{1\}$	$\{1\}$	$1-\theta$	0
6	$\{1\}$	$\{0\}$	$2\theta(1-\theta)$	0
7	$\{0, 1\}$	$\{0\}$	$\theta(2-\theta)$	0
8	$\{1\}$	$\{0, 1\}$	$1-\theta^2$	0

(ii) Binomial trials. We now develop example 4.5 (ii). This is such a simple case that we can enumerate all the possible predictors (table 5.1). We can then calculate the mean coverage for each of these eight predictors and these are also shown in table 5.1. We see immediately that the last five predictors are useless as mean coverage tolerance predictors since the infimum (5.1) associated with them is zero. The first three predictors have some possible use. We see that for $c \leqslant \frac{3}{4}$ we can obtain a tolerance predictor, for example δ_1, of mean coverage c, but that for $c > \frac{3}{4}$ no tolerance predictor of mean coverage c exists. Notice that we have omitted a decision rule of the form

$$\delta_9(0) = \{0, 1\}, \ \delta_9(1) = \{0, 1\}.$$

Although this would give mean coverage of 1 for all θ, it has no practical relevance and should not be considered as a tolerance predictor.

For this binomial trials situation no similar mean coverage tolerance predictor exists.

5.2 Interpretation of similar mean coverage tolerance predictors

Suppose that we wish to apply a predictor δ repeatedly on a number r of occasions on which nature may present us with different unknown indices according to an unknown density function $p(\theta)$. Then in the r applications of the predictor we may obtain $\theta_1, \ldots, \theta_r$, and information $x_1, \ldots, x_r$ in the r associated informative experiments. We would then use regions $\delta(x_1), \ldots, \delta(x_r)$, and the associated actual covers obtained would be $P_f\{\delta(x_1)|\theta_1\}, \ldots, P_f\{\delta(x_r)|\theta_r\}$. One criterion sometimes suggested is that the long-run average cover should attain some specified level c, usually not too far below 1: for large r

$$\frac{P_f\{\delta(x_1)|\theta_1\} + \ldots + P_f\{\delta(x_r)|\theta_r\}}{r} = c. \tag{5.5}$$

The counterpart of long-run average in a statistical model is the appropriate expectation, in this case with respect to the joint distribution of x and θ, namely $p(\theta)p_e(x|\theta)$. Thus we would seek a predictor δ satisfying

$$\mathrm{E}[P_f\{\delta(\cdot)|\cdot\}] = c \tag{5.6}$$

or

$$\int\int_{X \times \Theta} P_f\{\delta(x)|\theta\}p_e(x|\theta)p(\theta)\mathrm{d}x\mathrm{d}\theta = c. \tag{5.7}$$

We can express this double accumulation (integral or summation) as a repeated accumulation under very wide conditions:

$$\int_\Theta p(\theta)\left\{\int_X P_f\{\delta(x)|\theta\}\, p_e(x|\theta)\mathrm{d}x\right\}\mathrm{d}\theta = c. \tag{5.8}$$

If we can obtain a predictor which sets the inner accumulation equal to c for every θ, then the overall double accumulation will also be equal to c and condition (5.3) will be satisfied, whatever the nature of $p(\theta)$. This is the basis of definition 5.2 of a tolerance predictor of similar mean coverage c. Such a predictor has the property that if applied to a long run of similar informative experiments it provides long-run average cover c over the set of associated future experiments to which it is applied. Note that what is being said is not the following: if there is a single informative experiment with outcome x then the average cover provided by $\delta(x)$ over a long sequence of future experiments is c. The cover is in fact $P_f\{\delta(x)|\theta\}$ for each of these future experiments. Some such principle as that outlined above concerning what happens in a long run of repetitions of (e, f) has to be invoked to make use of mean coverage tolerance predictors.

There are other possible interpretations of the relation (5.3). For we may develop the integral in the following way:

$$\int_X P_f\{\delta(x)|\theta\}\, p_e(x|\theta)\mathrm{d}x = \int_X \left\{\int_{\delta(x)} p_f(y|\theta)\mathrm{d}y\right\} p_e(x|\theta)\mathrm{d}x$$

$$= P_{ef}\{(x,y) : y \in \delta(x)|\theta\}. \qquad (5.9)$$

Suppose that we can devise δ in such a way that this is equal to c for every $\theta \in \Theta$. Then in a long run of informative experiments each followed by a single prediction using the predictor δ the proportion of successful predictions (where the actual y observed falls in the corresponding prediction region $\delta(x)$) in the long run is c. There is an obvious extension to the form of (5.1).

5.3 Additional criteria

We have seen in example 5.1 (i) that a large number of tolerance predictors of mean coverage c may exist. The nature of the practical situation may, however, dictate just which of this class of possible tolerance predictors is a sensible one to use. For example the underlying density functions may describe the lifetime of a component from some production process. The manufacturer may wish to quote some minimum lifetime which he hopes will convey to the purchasers the quality of his product. To do this he will quote only prediction intervals of the form $(q_1 x, \infty)$. In our analysis above for the exponential distribution this would involve setting $q_2 = \infty$, and then there is a unique $q_1 = (1-c)/c$ leading to a given mean coverage c.

If the practical dictates of the situation do not lead to a unique predictor then some new principle has to be introduced to arrive at a satisfactory result. This may take the form of a principle of symmetry, the region having as its centre the most likely future observation or the mean future observation. We shall see how this principle is used in normal theory in §5.8. Some other principles depend on the notion that a large prediction region is less good than a smaller one. For instance, in example 5.1 (i), the length of the interval is $(q_2 - q_1)x$ so that the tolerance predictor of mean coverage c of the minimum length will minimise $q_2 - q_1$ subject to the condition (5.1). Since in the (q_1, q_2) plane of fig. 5.1 lines of the form

$$q_2 - q_1 = \text{constant}$$

each have unit gradient and since the gradient of the curve (5.4) everywhere exceeds 1 for $q_2 \geqslant q_1 \geqslant 0$ it is clear that the minimum occurs with $q_1 = 0$, $q_2 = c/(1-c)$.

One important principle of exclusion not yet mentioned is that of invariance. In example 5.1 (i) the lifetime of a component may be the variable under question. Suppose that the predictor δ specifies an interval $\{\delta_1(x), \delta_2(x)\}$. Since the prediction arrived at must be the same whether time is measured in minutes, microseconds or years we clearly require that

$$\delta_1(\lambda x) = \lambda\delta_1(x), \delta_2(\lambda x) = \lambda\delta_2(x) \quad (\lambda > 0); \tag{5.10}$$

the prediction procedure should be invariant under the group of transformations $x \to \lambda x$ (the group of changes of scale). Notice that the class of density functions already possesses the necessary property: if x is $\text{Ex}(\theta)$ then λx is $\text{Ex}(\theta/\lambda)$.

What limitation to the form of δ do the relations (5.10) imply? The answer to this question is easily seen since (5.10) are simply the defining relations of proportional functions $\delta_1(x) = q_1 x$, $\delta_2(x) = q_2 x$. We thus see that the restriction to invariant predictors in this case requires that we take the proportional function already considered in our analysis above. Thus it still leaves a wide class of possible tolerance predictors satisfying (5.4).

5.4 Relationship to Bayesian coverage

Mean coverage tolerance predictors have a strong Bayesian property which makes them particularly attractive. The following result provides this by pinpointing the relationship of the mean coverage of such a predictor to the concept of Bayesian coverage.

A tolerance predictor δ of similar mean coverage c has mean Bayesian coverage c whatever the prior density function $p(\theta)$ on Θ. (5.11)

The proof is as follows. Since δ has similar mean-coverage c we have, by (5.3),

$$\int_X P_f\{\delta(x)|\theta\}\, p_e(x|\theta)\mathrm{d}x = c \quad \text{for every} \quad \theta \in \Theta,$$

so that

$$\int_X \int_{\delta(x)} p_f(y|\theta)\, p_e(x|\theta)\mathrm{d}x\,\mathrm{d}y = c \quad \text{for every} \quad \theta \in \Theta.$$

Hence, for every prior density function $p(\theta)$ on Θ,

$$\int_\Theta \int_X \int_{\delta(x)} p_f(y|\theta)\, p_e(x|\theta)\, p(\theta)\mathrm{d}x\mathrm{d}y\mathrm{d}\theta = c. \tag{5.12}$$

Since, by (2.2),

$$p_e(x|\theta)p(\theta) = p(x)p(\theta|x)$$

we may rewrite (5.12) as

$$\int_X p(x) \int_{\delta(x)} \left\{ \int_\Theta p_f(y|\theta)\, p(\theta|x) d\theta \right\} dx dy = c$$

and so, by (2.4),

$$\int_X p(x) \int_{\delta(x)} p(y|x) dy dx = c. \tag{5.13}$$

The left side of (5.13) is the mean Bayesian coverage as defined in (4.15), and the result is therefore established.

The condition of similarity can be dropped from the statement (5.11) with a minor modification:

A tolerance predictor δ of mean coverage c has mean Bayesian coverage at least c whatever the prior density function p(θ) on Θ. (5.14)

5.5 Mean coverage tolerance predictors for the binomial case

Suppose that the informative experiment e is Bi(n, θ) and the future experiment f is Bi(N, θ). We consider first the construction of a tolerance predictor δ of mean coverage c and of the one-sided form

$$\delta(x) = \{0, 1, ..., \epsilon(x)\}. \tag{5.15}$$

Here the coverage (4.23) is

$$\sum_{y=0}^{\epsilon(x)} \binom{N}{y} \theta^y (1-\theta)^{N-y} \tag{5.16}$$

and has a distribution which is not independent of θ. This causes some difficulty in the construction of suitable $\epsilon(x)$. We must choose the region of summation (fig. 5.2) of the mean coverage, which is the double sum

$$\sum_{x=0}^{n} \sum_{y=0}^{\epsilon(x)} \binom{N}{y} \theta^y (1-\theta)^{N-y} \binom{n}{x} \theta^x (1-\theta)^{n-x}, \tag{5.17}$$

in such a way that (5.17) is at least c for every θ in the interval $[0, 1]$. Now the typical term of this double sum can be rewritten as

$$\binom{n+N}{x+y} \theta^{x+y} (1-\theta)^{n+N-(x+y)} \binom{n}{x} \binom{N}{y} \Big/ \binom{n+N}{x+y} \tag{5.18}$$

and the determination of $\epsilon(x)$ is made easier by a change in the method of summation. Consider the transformation

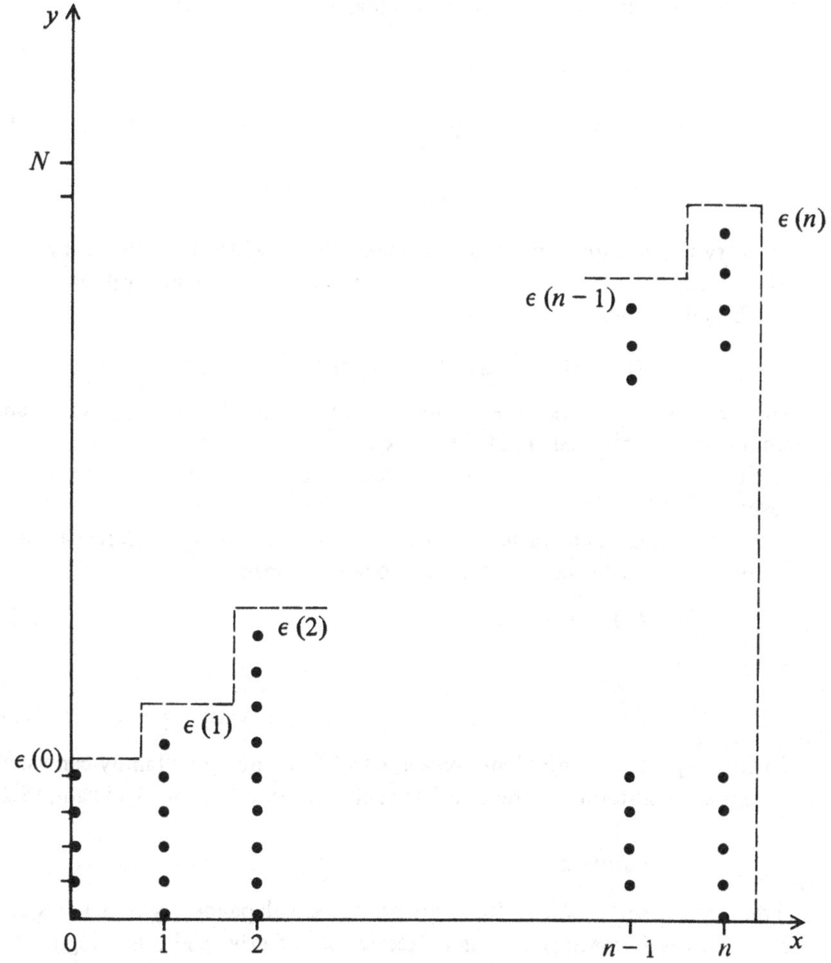

Fig. 5.2 Region of summation for the double sum associated with mean coverage (5.17).

$$u = x + y, \quad v = x. \tag{5.19}$$

The double sum can then be expressed as the repeated sum

$$\sum_{u=0}^{n+N} \binom{n+N}{u} \theta^u (1-\theta)^{n+N-u} \sum_{[v:v+\epsilon(v) \geq u]} \frac{\binom{n}{v}\binom{N}{u-v}}{\binom{n+N}{u}}. \tag{5.20}$$

Our object will then be achieved if we choose $\epsilon(v)$ such that

$$\sum_{[v:\, v+\epsilon(v) \geqslant u]} \frac{\binom{n}{v}\binom{N}{u-v}}{\binom{N+n}{u}} \geqslant c, \tag{5.21}$$

for every non-negative integer u. The inequality (5.21) will be satisfied in as economic a way as possible (that is with as small tolerance intervals as possible) if we take

$$\epsilon(x) = \min\{z: P_{\mathrm{hy}}(n+N, z, n, x) \leqslant 1-c\} - x - 1, \tag{5.22}$$

where P_{hy} is the hypergeometric function defined in (A21) of appendix I and tabulated by Lieberman and Owen (1961).

Notice that we cannot obtain a tolerance predictor δ of similar mean coverage c.

By an argument analogous to those above we can obtain a tolerance predictor of mean coverage c of the other one-sided form

$$\{\epsilon(x), \epsilon(x)+1, \ldots, N\} \tag{5.23}$$

by setting

$$\epsilon(x) = \max\{z: P_{\mathrm{hy}}(n+N, z, n, x-1) \geqslant c\} - x + 1. \tag{5.24}$$

Tolerance prediction problems associated with the binomial family can usually be expressed in terms of one or other of the one-sided forms (5.15) and (5.23).

Example 5.2

A quality control problem. Items are produced independently in batches of 30 by a firm. Items may be either effective or defective and it is recognised by both manufacturer and customer that batches vary considerably in the number of effectives they contain. The terms of a suggested contract between manufacturer and customer require the manufacturer to test destructively 5 of the 30 components from each batch, and to supply the remaining 25 components in a packet with an accompanying statement about the maximum number of defectives in the packet. The contract further requires that the statements must be correct for at least 80 per cent of batches. What statement strategy will fulfil the terms of the contract?

For a particular batch let θ denote the probability that an item will be defective, so that we can regard e and f as $\mathrm{Bi}(5, \theta)$ and $\mathrm{Bi}(25, \theta)$ experiments. If for an outcome x of e we state that the number y of defective items in the batch will be in the interval $\delta(x)$ then the contract requires that

$$P_{ef}\{(x, y)\colon\ y \in \delta(x) | \theta\} \geqslant 0.8 \tag{5.25}$$

whatever the value of θ. This is precisely the form (5.9) so that what we require is a mean coverage tolerance predictor of form (5.15) with $c = 0.8$, $n = 5, N = 25$. Consider the construction of the tolerance prediction corresponding to $x = 1$. The hypergeometric function within the braces of (5.22) is then

$$P_{\mathbf{hy}}(30, z, 5, 1).$$

From the Lieberman and Owen (1961) tables we have

$$P_{\mathbf{hy}}(30, 15, 5, 1) = 0.1648, \quad P_{\mathbf{hy}}(30, 14, 5, 1) = 0.2095,$$

so that, by (5.22),

$$\epsilon(1) = 15 - 1 - 1 = 13.$$

The complete mean coverage tolerance predictor is set out in table 5.2.

Table 5.2 *Mean coverage predictor for the quality control problem*
$\delta(x) = \{0, 1, ..., \epsilon(x)\}$

x	0	1	2	3	4	5
$\epsilon(x)$	7	13	17	21	24	25

5.6 Mean coverage tolerance predictors for the Poisson case

Consider the construction of a tolerance predictor δ of the form

$$\delta(x) = \{0, 1, ..., \epsilon(x)\} \tag{5.26}$$

for a future $\mathrm{Po}(K\theta)$ experiment f, based on the information from a $\mathrm{Po}(k\theta)$ informative experiment e and with mean coverage c. Similarly to the binomial case we have to ensure that the inequality

$$\sum_{x=0}^{\infty} \sum_{y=0}^{\epsilon(x)} \frac{\exp(-K\theta)(K\theta)^y}{y!} \frac{\exp(-k\theta)(k\theta)^x}{x!} \geqslant c \tag{5.27}$$

holds for every $\theta > 0$. Again we consider an alternative way of carrying out the double summation. The double sum may be written as

$$\Sigma\Sigma \binom{x+y}{x} \left(\frac{k}{k+K}\right)^x \left(\frac{K}{k+K}\right)^y \frac{\exp\{-(k+K)\theta\}\{(k+K)\theta\}^{x+y}}{(x+y)!} \tag{5.28}$$

where the region of summation is again as shown in fig. 5.2 with $n \to \infty, N \to \infty$. Again the change of variables

$$u = x + y, \quad v = x$$

gives

$$\sum_{u=0}^{\infty} \frac{\exp\{-(k+K)\theta\}\{(k+K)\theta\}^u}{u!}$$

$$\times \sum_{[v:\, v+\epsilon(v) \geqslant u]} \binom{u}{v} \left(\frac{k}{k+K}\right)^v \left(\frac{K}{k+K}\right)^{u-v}. \tag{5.29}$$

The inequalities analogous to (5.21) then lead us to set

$$\epsilon(x) = \min\{z: I_{k/(k+K)}(x+1, z) \geqslant c\} - 1, \tag{5.30}$$

where I is the incomplete beta function defined in (A19) of Appendix I and tabulated by Pearson (1934).

For a predictor of the form

$$\delta(x) = \{\epsilon(x), \epsilon(x)+1, ...,\} \tag{5.31}$$

and with mean coverage c we set

$$\epsilon(x) = \max\{z: I_{k/(k+K)}(x, z) \leqslant 1 - c\}. \tag{5.32}$$

Example 5.3

Seed germination. A rare plant bears many seeds of low germination rate. A seedsman divides each of his boxes of 12 such plants at random into two sets of 6, sows the seeds from one set for his own crops next season, records the number of germinating seeds from the set, and packets the seeds from the other set with a statement about the minimum number of seeds from the packet likely to germinate. If he hopes that 95 per cent of his packets should contain correct statements what minimum number of germinating seeds should he attribute to a packet associated with 6 plants yielding 7, 3, 5, 5, 4, 6 germinating seeds?

Let us assume that the number of germinating seeds from a plant is distributed as $\text{Po}(\theta)$, where θ may vary from box to box. The informative experiment e is then effectively $\text{Po}(6\theta)$ with observation $x = 7 + 3 + ... + 6 = 30$; the future experiment f, which records the number y of germinating seeds, is also $\text{Po}(6\theta)$. To achieve 95 per cent statements we require to quote minima $\epsilon(x)$ satisfying

$$P_{ef}\{(x, y):\ y \geqslant \epsilon(x) | \theta\} \geqslant 0.95 \tag{5.33}$$

for every $\theta > 0$. This is of form (5.9) and we are thus led to seek a mean coverage tolerance predictor of type (5.31) with $k = K = 6$, $c = 0.95$. The incomplete beta function within the braces of (5.32) is, for $x = 30$,

$$I_{0.5}(30, z).$$

Since

$$I_{0.5}(30, 18) = 0.0395, \quad I_{0.5}(30, 19) = 0.0557,$$

we have

$$\epsilon(30) = 18.$$

5.7 Mean coverage tolerance predictors for the gamma case

In example 5.1 (i) we saw that it is possible to construct a tolerance predictor of similar mean coverage c for the simple exponential case. This particular case can be included as a special case of the following more general formulation. Suppose that

$$p_e(x|\theta) = \text{Ga}(k, \theta), \; p_f(y|\theta) = \text{Ga}(K, \theta).$$

For the reasons of invariance suggested in §5.3 it is natural to consider a predictor of the form

$$\delta(x) = (q_1 x, q_2 x). \tag{5.34}$$

We can obtain from this finite-interval predictor one-sided predictors by considering the cases $q_1 = 0$ and $q_2 = \infty$ as special forms.

The coverage of δ is

$$\int_{q_1 x}^{q_2 x} \frac{\theta^K y^{K-1} \exp(-\theta y)}{\Gamma(K)} \, dy \tag{5.35}$$

which has a distribution independent of θ, so that in mean coverage calculations we can conveniently set $\theta = 1$. The mean coverage is then given by (5.9) as

$$P_{ef}\{(x, y): q_1 x \leqslant y \leqslant q_2 x\},$$

where x, y are independent Ga$(k, 1)$ and Ga$(K, 1)$ random variables. This mean coverage can thus be expressed as

$$P_{ef}\left\{(x, y): \frac{1}{1+q_2} \leqslant \frac{x}{x+y} \leqslant \frac{1}{1+q_1}\right\}. \tag{5.36}$$

Since $x/(x+y)$ has a Be(k, K) distribution, we see that we obtain mean coverage c if we choose c_1, c_2 such that $0 \leqslant c_2 \leqslant c_1 \leqslant 1$ and $c_1 - c_2 = c$ and then set, in the quantile notation of appendix I,

$$\frac{1}{1+q_1} = \text{Be}(k, K; c_1), \tag{5.37}$$

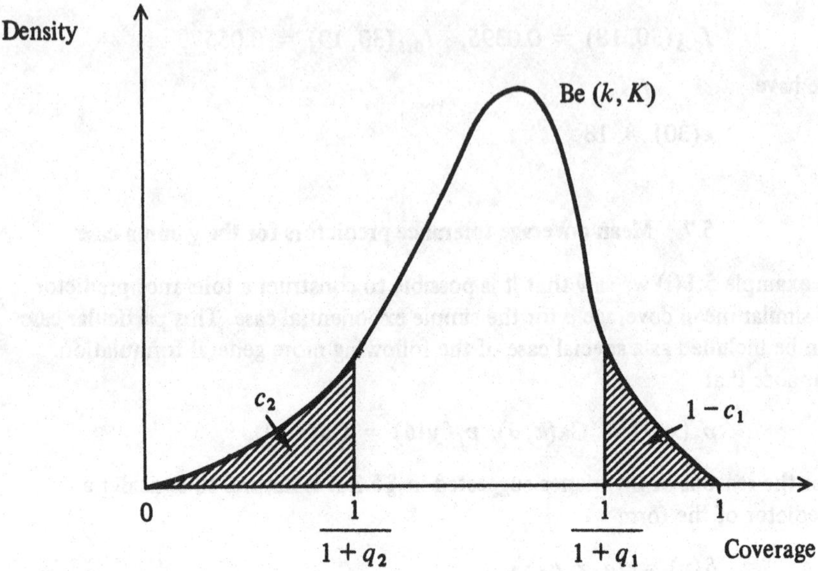

Fig. 5.3 Mean coverage predictors for the gamma case given by (5.37), (5.38).

$$\frac{1}{1+q_2} = \text{Be}(k, K; c_2); \tag{5.38}$$

see fig. 5.3. Thus we obtain the following general result for this case. The tolerance predictor

$$\delta(x) = \left\{ \frac{1-\text{Be}(k, K; c_1)}{\text{Be}(k, K; c_1)}x, \ \frac{1-\text{Be}(k, K; c_2)}{\text{Be}(k, K; c_2)}x \right\} \tag{5.39}$$

has similar mean coverage c.

The following semi-infinite interval predictors are derived as special cases. The tolerance predictors

$$\delta(x) = \left\{ 0, \frac{1-\text{Be}(k, K; 1-c)}{\text{Be}(k, K; 1-c)}x \right\} \tag{5.40}$$

and

$$\delta(x) = \left\{ \frac{1-\text{Be}(k, K; c)}{\text{Be}(k, K; c)}x, \infty \right\} \tag{5.41}$$

each have similar mean coverage c.

5.8 Mean coverage tolerance predictors for the normal case

We suppose throughout this section that

$$p_f(y\,|\,\theta) = \text{No}(\mu, K\tau)$$

and that the informative experiment e provides (m, v), where m and v are independent with $\text{No}(\mu, k\tau)$ and $\text{Ch}(\nu, \tau)$ distributions.

For example we may take

$$m = \bar{x} = \sum x_i/n, \; v = S(\mathbf{x}, \mathbf{x}),$$

$$k = n, \nu = n-1,$$

for the simple normal prediction problem where e consists of n replicates of a $\text{No}(\mu, \tau)$ experiment. We find it more convenient to use v than the conventional standard deviation statistic s given by $s = (v/\nu)^{1/2}$.

One-sided tolerance predictors. We consider here a predictor δ defined by

$$\delta(m, v) = (-\infty, m + qv^{1/2}), \tag{5.42}$$

where q is a constant to be determined. The choice of the form $m + qv^{1/2}$ is based on an invariance property, that the predictor should be invariant under the group of transformations $m \to m + g$, $v \to hv\,(h > 0)$. The coverage of δ is

$$\Phi\{(K\tau)^{1/2}(m + qv^{1/2} - \mu)\} \tag{5.43}$$

and has a distribution which is clearly independent of $\theta = (\mu, \tau)$. In computing the mean coverage we can thus simplify matters by setting $(\mu, \tau) = (0, 1)$. If therefore we use the abbreviated notation,

$$p(m) = \text{No}(0, k), \; p(v) = \text{Ch}(\nu, 1), \; p(y) = \text{No}(0, K), \tag{5.44}$$

we can express the mean coverage as

$$\int_{-\infty}^{\infty}\int_0^{\infty} \Phi\{K^{1/2}(m + qv^{1/2})\}\, p(m)\, p(v) dm\, dv$$

$$= P_{ef}\{(m, v, y):\; y \leqslant m + qv^{1/2}\}$$

$$= P_{ef}\left[(m, v, y): (y-m)\Big/\left\{\frac{v}{\nu}\left(\frac{1}{k}+\frac{1}{K}\right)\right\}^{1/2} \leqslant q\Big/\left\{\frac{1}{\nu}\left(\frac{1}{k}+\frac{1}{K}\right)\right\}^{1/2}\right]. \tag{5.45}$$

Now

$$(y-m)\Big/\left\{\frac{v}{\nu}\left(\frac{1}{k}+\frac{1}{K}\right)\right\}^{1/2}$$

is distributed as $t(\nu)$ and so to achieve mean coverage c (for every θ) we may set

$$q\Big/\left\{\frac{1}{\nu}\left(\frac{1}{k}+\frac{1}{K}\right)\right\}^{1/2} = t(\nu;c).$$

Thus we have the following result.

The tolerance predictor

$$\delta(m,v) = \left[-\infty, m+\left\{\frac{1}{\nu}\left(\frac{1}{k}+\frac{1}{K}\right)\right\}^{1/2} t(\nu;c)v^{1/2}\right] \tag{5.46}$$

has similar mean coverage c.

In exactly the same manner it can be shown that

$$\delta(m,v) = \left[m-\left\{\frac{1}{\nu}\left(\frac{1}{k}+\frac{1}{K}\right)\right\}^{1/2} t(\nu;c)v^{1/2}, \infty\right] \tag{5.47}$$

has similar mean coverage c.

Symmetric two-sided tolerance predictors. The prediction intervals obtained in (5.46) and (5.47) are semi-infinite, and this may not suit the particular practical problem. Where it is desirable to quote a finite interval a popular tolerance predictor is one that is symmetric about m, namely of the form

$$\delta(m,v) = (m-qv^{1/2}, m+qv^{1/2}). \tag{5.48}$$

The coverage,

$$\Phi\{(K\tau)^{1/2}(m+qv^{1/2}-\mu)\}-\Phi\{(K\tau)^{1/2}(m-qv^{1/2}-\mu)\}, \tag{5.49}$$

has again a distribution which is independent of (μ,τ) and the same device of considering the special case $(\mu,\tau)=(0,1)$ may be used in evaluating the mean coverage:

$$\int_{-\infty}^{\infty}\int_{0}^{\infty}[\Phi\{K^{1/2}(m+qv^{1/2})\}-\Phi\{K^{1/2}(m-qv^{1/2})\}]\,p(m)p(v)dmdv$$

$$= P_{ef}\{(m,v,y):\ m-qv^{1/2}\leqslant y\leqslant m+qv^{1/2}\}. \tag{5.50}$$

Again the fact that $(y-m)\Big/\left\{\frac{v}{\nu}\left(\frac{1}{k}+\frac{1}{K}\right)\right\}^{1/2}$ is distributed as $t(\nu)$ gives an immediate determination of

$$q = \left\{\frac{1}{\nu}\left(\frac{1}{k}+\frac{1}{K}\right)\right\}^{1/2} t\{\nu;\tfrac{1}{2}(c+1)\}$$

yielding the following result.

The tolerance predictor

$$\left[m - \left\{\frac{1}{\nu}\left(\frac{1}{k} + \frac{1}{K}\right)\right\}^{1/2} t\{\nu; \tfrac{1}{2}(1+c)\}v^{1/2}, m + \left\{\frac{1}{\nu}\left(\frac{1}{k} + \frac{1}{K}\right)\right\}^{1/2} t\{\nu; \tfrac{1}{2}(1+c)\}v^{1/2}\right] \tag{5.51}$$

has similar mean coverage c.

Non-symmetric two-sided tolerance predictors. It is clear from the construction of (5.51) that it is easy to construct a tolerance predictor of mean coverage c of the form

$$\delta(m, v) = (m + q_1 v^{1/2}, m + q_2 v^{1/2}), \tag{5.52}$$

where q_1 is not necessarily equal to $-q_2$. This is achieved by setting

$$q_1 = \left\{\frac{1}{\nu}\left(\frac{1}{k} + \frac{1}{K}\right)\right\}^{1/2} t(\nu; c_1), \tag{5.53}$$

$$q_2 = \left\{\frac{1}{\nu}\left(\frac{1}{k} + \frac{1}{K}\right)\right\}^{1/2} t(\nu; c_2), \tag{5.54}$$

where $0 \leqslant c_1 < c_2 < 1$ and $c_2 - c_1 = c$. The symmetric case is given by $c_2 = \frac{1}{2}(1+c), c_1 = \frac{1}{2}(1-c)$, but there are many other (c_1, c_2) combinations.

The length of the interval provided by δ depends on the magnitude of $q_2 - q_1$ and it can be shown that this is a minimum for fixed c when $q_1 = q_2$, the symmetric case.

Example 5.4

Crop prediction. A potato farmer wishes to gauge the weight of crop he is likely to produce from the potatoes he has planted in separate fields so that he may provide relevant information to prospective buyers. A few days before complete harvesting he selects at random 24 square metre units in a field and measures the weight of crop produced in each unit. He then wishes to provide a 95 per cent mean coverage tolerance interval for the yield per square metre of the crop. (This can easily be converted to an interval for the total yield in the field.)

We assume that the weights in each square metre are independent $\text{No}(\mu, \tau)$ where (μ, τ) may vary from field to field. The informative experiment e thus effectively provides independent m and v where m is $\text{No}(\mu, 24\tau)$ and v is $\text{Ch}(23, \tau)$. Suppose the values obtained for the 24 measurements are as in table 5.3. Then $m = 7.97$ and $v = 57.87$. Hence substitution in (5.51) gives a tolerance interval of similar mean coverage 0.95 of the form (4.62, 11.32). In the same way we can obtain one sided intervals $(-\infty, 10.74)$ or $(5.19, \infty)$ of similar mean coverage 0.95 by substitution in (5.46) or (5.47).

Table 5.3 *Potato crop yields (kg) for 24 square metre units*

10.8	8.0	7.3	11.4	7.4	7.0	4.1	7.1
8.3	7.7	9.3	7.5	9.7	8.3	7.1	9.8
9.3	8.7	7.4	7.8	7.3	5.9	6.0	8.0

5.9 Mean coverage tolerance predictors for the multinormal case

It is relatively easy to obtain a similar mean coverage tolerance predictor for the multinormal case analogous to (5.48) for the normal case. Suppose that

$$p_f(\mathbf{y}\,|\,\boldsymbol{\theta}) = \mathrm{No}_d(\boldsymbol{\mu}, K\boldsymbol{\tau}) \tag{5.55}$$

and that the informative experiment e provides $(\mathbf{m}, \mathbf{v})$, where $\mathbf{m}$ and $\mathbf{v}$ are independently distributed as $\mathrm{No}_d(\boldsymbol{\mu}, k\boldsymbol{\tau})$ and $\mathrm{Wi}_d(\nu, \boldsymbol{\tau})$. We consider a predictor δ giving predictive regions which are ellipsoidal in shape:

$$\delta(\mathbf{m}, \mathbf{v}) = \{\mathbf{y}: (\mathbf{y}-\mathbf{m})'\mathbf{v}^{-1}(\mathbf{y}-\mathbf{m}) \leqslant q\}, \tag{5.56}$$

the counterpart of (5.48). The problem is to determine q such that δ has mean coverage c. By (5.9) the mean coverage can be expressed as

$$P_{ef}\{(\mathbf{m}, \mathbf{v}, \mathbf{y}): (\mathbf{y}-\mathbf{m})'\mathbf{v}^{-1}(\mathbf{y}-\mathbf{m}) \leqslant q\,|\,\boldsymbol{\theta}\}. \tag{5.57}$$

It is again easily verified that the value of this is independent of $\boldsymbol{\theta} = (\boldsymbol{\mu}, \boldsymbol{\tau})$, and so similar mean coverage tolerance predictors can be constructed. For simplicity of evaluation we can thus set $(\boldsymbol{\mu}, \boldsymbol{\tau}) = (\mathbf{0}, \mathbf{I}_d)$, for which case $\mathbf{y}-\mathbf{m}$ is

$$\mathrm{No}_d\left\{\mathbf{0}, \left(\frac{1}{k}+\frac{1}{K}\right)^{-1}\mathbf{I}_d\right\}$$

independently of $\mathbf{v}$, which is $\mathrm{Wi}_d(\nu, \mathbf{I}_d)$.

Two routes to the determination of q are available. The first shows that the quadratic form

$$\nu(\mathbf{y}-\mathbf{m})'\mathbf{v}^{-1}(\mathbf{y}-\mathbf{m}) \Big/ \left(\frac{1}{k}+\frac{1}{K}\right)$$

has a T^2 distribution (Anderson, 1958), which through its relation to the F distribution, yields

$$q = \left(\frac{1}{k}+\frac{1}{K}\right)\frac{d}{\nu-d+1}\,\mathrm{F}(d, \nu-d+1; c). \tag{5.58}$$

The other exploits the property (4.12) by observing that if $\mathbf{W}'\mathbf{W} = \mathbf{v}$ then

$$\mathbf{z} = \nu^{1/2}\mathbf{W}^{-1}(\mathbf{y}-\mathbf{m}) \Big/ \left(\frac{1}{k}+\frac{1}{K}\right)^{1/2} \quad \text{is} \quad \mathrm{St}_d(\nu, \mathbf{0}, \mathbf{I}_d).$$

Hence (5.57) can be expressed as

$$P\left\{z: z'z \leqslant \nu q \Big/ \left(\frac{1}{k}+\frac{1}{K}\right)\right\},$$

which by (4.12) can be evaluated as

$$I_{r/(r+1)}\{\tfrac{1}{2}d, \tfrac{1}{2}(\nu-d+1)\},$$

where

$$r = q\Big/\left(\frac{1}{k}+\frac{1}{K}\right).$$

Hence

$$q = \left(\frac{1}{k}+\frac{1}{K}\right)\frac{\mathrm{Be}\{\tfrac{1}{2}d, \tfrac{1}{2}(\nu-d+1); c\}}{1-\mathrm{Be}\{\tfrac{1}{2}d, \tfrac{1}{2}(\nu-d+1); c\}}, \tag{5.59}$$

which is equivalent to (5.58).

5.10 Mean coverage tolerance predictors for the two-parameter exponential case

Suppose that for the future experiment

$$p_f(y\,|\,\theta) = \mathrm{Er}(\mu, K\tau) \tag{5.60}$$

and that the information from e can be condensed to (m, v), where m and v are independent with density functions

$$p(m\,|\,\mu,\tau) = \mathrm{Er}(\mu, k\tau), \quad p(v\,|\,\tau) = \mathrm{Ga}(\nu, \tau). \tag{5.61}$$

One-sided tolerance predictor. We consider a predictor of the form

$$\delta(m, v) = (m+qv, \infty) \tag{5.62}$$

where $q \geqslant 0$. Since $m \geqslant \mu$ and $v \geqslant 0$ the restriction to the case $q \geqslant 0$ ensures that $m+qv \geqslant \mu$ and avoids a complication of the coverage distribution which otherwise arises. The coverage of the predictor (5.62) is

$$\exp\{-K\tau(m+qv-\mu)\} \tag{5.63}$$

which has a distribution independent of $\theta = (\mu, \tau)$. Thus in computing mean coverage we can consider the convenient case $(\mu, \tau) = (0, 1)$. The mean coverage is

$$\int_0^\infty \int_0^\infty \exp(-Km - qKv)p(m)\,p(v)\,\mathrm{d}m\,\mathrm{d}v$$

$$= \int_0^\infty \exp(-Km)\,p(m)\,\mathrm{d}m \int_0^\infty \exp(-qKv)\,p(v)\,\mathrm{d}v \tag{5.64}$$

where

$$p(m) = \text{Er}(0, k) = \text{Ex}(k),$$
$$p(v) = \text{Ga}(\nu, 1).$$

Integration of (5.64) now gives mean coverage as

$$\frac{k}{k+K}\,\frac{1}{(1+qK)^{\nu}} \tag{5.65}$$

with the following conclusion.

The tolerance predictor

$$\delta(m, v) = \left[m + \frac{v}{K}\left\{\left(\frac{k}{c(k+K)}\right)^{1/\nu} - 1\right\}, \infty\right] \tag{5.66}$$

has similar mean coverage c. Note that a condition for the existence of such a predictor with $q \geqslant 0$ is that

$$\frac{k}{c(k+K)} \geqslant 1,$$

that is

$$k \geqslant cK/(1-c). \tag{5.67}$$

The critical quantity $k/(k+K)$ involved in this condition is, of course, simply the mean coverage of the tolerance predictor $\delta(m, v) = (m, \infty)$.

The relaxing of the condition that $q \geqslant 0$ alters expression (5.63) for the coverage of (5.62) to

$$\begin{cases} \exp\{-K\tau(m+qv-\mu)\} & \text{if } m+qv \geqslant \mu, \\ 1 & \text{if } m+qv < \mu, \end{cases} \tag{5.68}$$

and the ensuing mean coverage computation is more involved. This complication can be avoided by ensuring that the informative experiment is large enough for (5.67) to hold.

Two-sided tolerance predictor. As an example of a tolerance predictor providing a finite interval we consider

$$\delta(m, v) = (m, m+qv) \tag{5.69}$$

with $q \geqslant 0$. The technique of evaluating mean coverage is now so familiar that we omit details here and quote the conclusion.

The tolerance predictor

$$\delta(m, v) = \left[m, m + \frac{v}{K}\left\{\left(1 - \frac{c(k+K)}{k}\right)^{-(1/\nu)} - 1\right\}\right] \tag{5.70}$$

has similar mean coverage c. Note again the condition $k > cK/(1-c)$ for the existence of such a predictor.

History

The concept of mean coverage tolerance predictors was introduced by Wilks (1941) who constructed such predictors for the normal case. For the multinormal case, see Chew (1966) and Guttman (1970). Most writers appear to have concentrated on the normal case. For some of the other cases, see Aitchison and Sculthorpe (1965). See also Bain and Weeks (1965), Paulson (1943) and Proschan (1953).

Problems

5.1 Suppose that the informative experiment e is $\text{Ge}(\theta)$, with density function

$$p_e(x|\theta) = \theta^x(1-\theta) \quad (x = 0, 1, 2, \ldots),$$

where $0 < \theta < 1$, and that the future experiment f is $\text{Ge}(\theta)$. Derive tolerance predictors of mean coverage c of the form

(i) $\delta(x) = \{0, 1, \ldots, \epsilon(x)\}$,

(ii) $\delta(x) = \{\epsilon(x), \epsilon(x)+1, \ldots\}$.

5.2 Suppose that the future experiment f is described by a Pareto distribution $\text{Pa}(k, \theta)$, with density function

$$p_f(y|\theta) = \frac{\theta k^\theta}{x^{\theta+1}} \quad (x > k),$$

where $\theta > 0$ and k is a known constant, and that the informative experiment provides $x_1, x_2, \ldots, x_n$ from n replicates of $\text{Pa}(k, \theta)$. Show that

$$v = \log\left(\frac{x_1 x_2 \ldots x_n}{k^n}\right)$$

is sufficient for θ with $p_e(v|\theta) = \text{Ga}(n, \theta)$. Derive mean coverage tolerance predictors of the forms

(i) $\delta(v) = (k + qv, \infty)$,

(ii) $\delta(v) = (k, k + qv)$,

(iii) $\delta(v) = (k + q_1 v, k + q_2 v)$.

5.3 After the completion of the first three out of each set of eight binomial trials, each with outcome either response or non-response, an attempt is to be made to estimate the minimum number of responses that will be obtained in the complete set. This estimation attempt is to be made for each of a long series of such sets, and it is hoped that about 80 per cent of the attempts will be successful. Provide a suitable procedure to attain this target.

5.4 Standard bales of cloth of length 50 metres have 'blemishes' distributed along their length according to a Poisson process. For each bale of cloth the number of blemishes is counted along 10 metres chosen at random and then a label is attached to the bale stating the estimated maximum number of blemishes for that bale. The manufacturer wants the information on 95 per cent of these labels to be correct.

What maximum number of blemishes should be quoted for standard bales whose sampled 10 metres have shown (i) 1 blemish, (ii) 7 blemishes?

5.5 Light bulbs are produced independently in batches of 50. The lifetimes of bulbs in a batch may be assumed to be described by an $\text{Ex}(\theta)$ distribution, but θ may vary from batch to batch. Before marketing a batch the manufacturer wishes to make some statement about the minimum lifetime of a bulb from the batch, and to be correct in his statements 95 per cent of the time. He therefore selects 5 bulbs at random from the batch and measures their lifetimes simultaneously. What statement should he make if the lifetimes (in hours) are 111, 86, 110, 50, 16?

Suppose that the manufacturer decides that he cannot afford to spend more than 72 hours on testing the bulbs from any batch. He therefore counts the number out of the 5 which have failed in that time. Can you help him now to make the relevant statement about the minimum lifetime of a bulb in the remainder of the batch?

5.6 Consider again the 'self-destructive' components described in problem 4.12. These have been designed so that each lifetime is described by an $\text{Er}(\mu, \tau)$ random variable. Suppose, however, that we are unable to specify how μ, τ vary over production runs. Each run produces 5 components. It is so important to the customer to have an accurate description of the lifetime that he persuades the manufacturer to test 4 of the 5 components and then provide him with a statement about the minimum lifetime of the remaining component. The manufacturer realises that there is not much evidence available and says that he can only be correct in his statements 75 per cent of the time. What minimum lifetime should he report if the 4 tested components had lifetimes 306.5, 301.1, 304.2, 320.3 mins?

5.7 After completing a standard linear regression analysis in which you have analysed the relationship of a response to a stimulus you are asked to express a view as to what response will occur if a stimulus of strength t is applied to a new experimental unit.

Can you state an interval within which you can claim 'with 95 per cent confidence' that the new response will lie?

Apply the theory you have developed to the following data to make a statement concerning the use of the stimulus at strength 1.5.

Stimulus strength	1	2	3
Response	1	4	4

5.8 In the analysis of the metabolic excretion rates in §4.4 a region of previous experience was constructed. This had the property that the probability of a new hyperplasia patient having metabolic rates within the region is assessed (in terms of the predictive distribution) to be 0.99. For this set of data construct an elliptical tolerance prediction of mean coverage 0.99, and compare this with the region of previous experience.

5.9 Show that the region of previous experience based on independently distributed $\mathrm{No}_d(\boldsymbol{\mu}, \boldsymbol{\tau})$ vectors $\mathbf{x}_1, \ldots, \mathbf{x}_n$ is a similar mean coverage tolerance predictor with mean coverage c, where c is given by

$$F(d, n-d; c) = \frac{n(n-d)}{d(n+1)} \max_i (\mathbf{x}_i - \bar{\mathbf{x}})' \mathbf{v}^{-1} (\mathbf{x}_i - \bar{\mathbf{x}}),$$

where

$$\bar{\mathbf{x}} = \frac{1}{n}(\mathbf{x}_1 + \ldots + \mathbf{x}_n), \quad \mathbf{v} = \sum_i (\mathbf{x}_i - \bar{\mathbf{x}})(\mathbf{x}_i - \bar{\mathbf{x}})'.$$

6
Guaranteed coverage tolerance prediction

6.1 Introduction

We saw in §4.6 that the mathematical achievement of a guaranteed coverage tolerance predictor is to ensure that the bulk of its coverage distribution lies above a specified value, conveniently termed the guaranteed coverage, say c. To make this idea concrete we had to specify what we meant by 'bulk' and we chose a definition involving quantiles of the coverage distributions. This approach can be conveniently expressed in terms of the concept of the guarantee of a predictor. The first step is to define the guarantee function of a predictor δ in providing cover c; this is a function of θ, the indexing parameter (cf. the concept of power function). We define $g(\theta\,|\,\delta, c)$ to be the probability that δ produces a region with cover of at least c, that is

$$g(\theta\,|\,\delta, c) = P_e[x : P_f\{\delta(x)|\theta\} \geqslant c\,|\,\theta]. \tag{6.1}$$

This quantity, for given θ, is simply the amount of the coverage distribution to the right of c. We can then define the guarantee of δ in providing coverage c.

Definition 6.1

Guarantee provided by a predictor. The ***guarantee*** of coverage c provided by a predictor δ is

$$g(\delta, c) = \inf_\Theta g(\theta\,|\,\delta, c). \tag{6.2}$$

The formal definition of a guaranteed coverage tolerance predictor is then immediate.

Definition 6.2

Guaranteed coverage tolerance predictor. A (c, g) ***guaranteed coverage tolerance predictor*** δ provides coverage c with guarantee g if $g(\delta, c) = g$, that is if

$$\inf_\Theta P_e[x : P_f\{\delta(x)|\theta\} \geqslant c\,|\,\theta] = g. \tag{6.3}$$

For brevity, we shall refer to such a predictor as a (c, g) tolerance predictor. For the case of discrete distributions the equality sign in definition 6.2 is relaxed to $\geqslant$.

It can happen that the guarantee function is constant, that is, does not depend on θ. For such predictors the familiar terminology of similarity is used, thus giving the following definition.

Definition 6.3

Similar guaranteed coverage tolerance predictor. A tolerance predictor δ has ***similar guaranteed coverage*** c with guarantee g if

$$g(\theta \mid \delta, c) = g \text{ for every } \theta \in \Theta. \tag{6.4}$$

Note that as in the case of mean coverage tolerance predictors we have again extended the usual definitions and that our similar guaranteed coverage tolerance predictors are the usual frequentist guaranteed coverage predictors.

Example 6.1

(i) *Exponential distribution.* We have already seen in example 4.5(i) that a (c, g) tolerance predictor exists. Here we rework the example in terms of the guarantee function. We have, for $\delta(x) = (qx, \infty)$,

$$\begin{aligned} g(\theta \mid \delta, c) &= P_e\{x : \exp(-q\theta x) \geqslant c \mid \theta\} \\ &= P_e\left\{x : x \leqslant -\frac{1}{q\theta} \log c \mid \theta\right\} \\ &= 1 - c^{1/q}, \end{aligned} \tag{6.5}$$

which is independent of θ, so that

$$g(\delta, c) = 1 - c^{1/q}, \tag{6.6}$$

and we obtain a similar (c, g) tolerance predictor by setting

$$1 - c^{1/q} = g,$$

that is,

$$q = \frac{\log c}{\log(1 - g)}, \tag{6.7}$$

as before.

That the problem is not always trivial can be seen if we consider a predictor $\delta(x) = (q_1 x, q_2 x)$ which provides finite intervals. Then

$$g(\theta \mid \delta, c) = P_e\{x : \exp(-q_1\theta x) - \exp(-q_2\theta x) \geqslant c \mid \theta\}. \tag{6.8}$$

It is easy to see that this is independent of θ and so $g(\delta, c)$ is the value of $g(\theta \mid \delta, c)$ at any convenient value of θ, say $\theta = 1$. But the evaluation of

$$g(\delta, c) = \int_{[x:\, \exp(-q_1 x) - \exp(-q_2 x) > c]} \exp(-x)\, dx \tag{6.9}$$

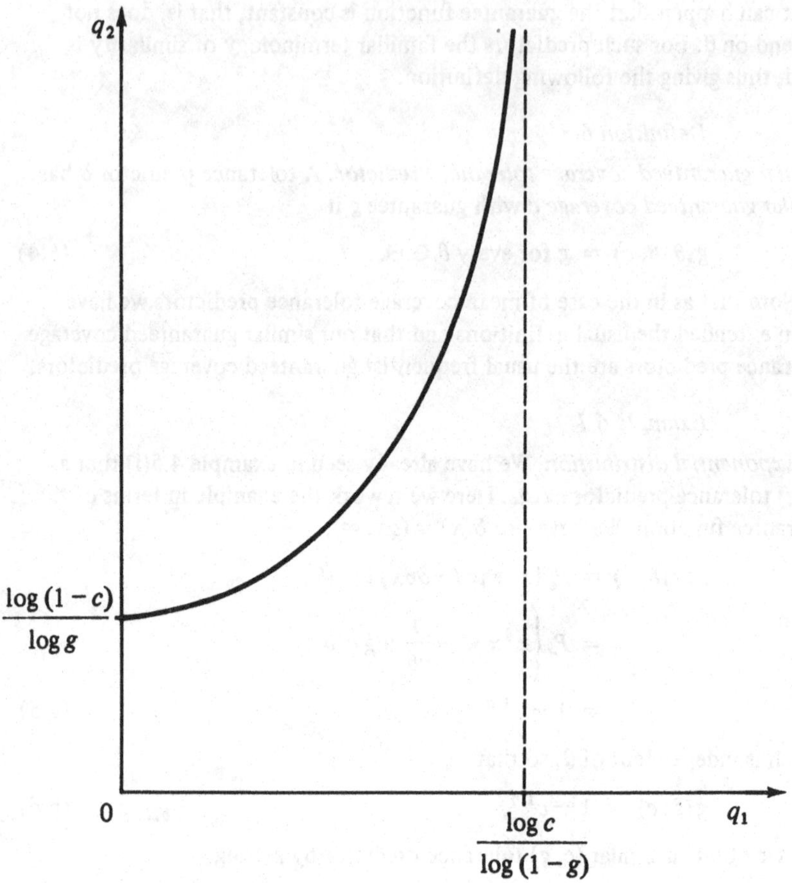

Fig. 6.1 Locus of (q_1, q_2) satisfying (6.9) for a guaranteed coverage tolerance prediction in the exponential case.

and the solution of $g(\delta, c) = g$ can be achieved only by numerical methods. Indeed there is a whole infinity of possible pairs (q_1, q_2), lying in the (q_1, q_2) plane along a continuous curve (fig. 6.1) joining $\{0, \log(1-c)/\log g\}$ to $\{\log c/\log(1-g), \infty\}$, the latter corresponding to the (c, g) tolerance predictor of the form $(q_1 x, \infty)$ which we have just been investigating.

(ii) *Binomial trials*. The guarantee function for each of the eight predictors of example 5.1(ii) can readily be constructed. For example, for $c > \frac{1}{2}$ which would normally be a practical requirement,

$$g(\theta \mid \delta_3, c) = \begin{cases} 1-\theta & (\theta \leqslant 1-c), \\ 0 & (1-c < \theta < c), \\ \theta & (\theta \geqslant c), \end{cases} \tag{6.10}$$

and so

$$g(\delta_3, c) = 0. \tag{6.11}$$

The guarantee is also zero for predictors $\delta_4, \ldots, \delta_8$ of table 5.1. However

$$g(\delta_1, c) = g(\delta_2, c) = 1 - c,$$

and hence (c, g) tolerance predictors exist for this case only if $1 - c \geqslant g$.

6.2 Interpretation of similar guaranteed coverage tolerance predictors

Users of guaranteed coverage predictors are often unclear about their suitability and it is therefore important to give specific interpretations of their properties. One such interpretation is to relate the concepts of guaranteed coverage c and guarantee g to the performance of the predictor δ in a series of repeated applications, one following each of a series of informative experiments. Suppose that we regard as successful (although in practice we may have very limited means of assessing success) any particular prediction that provides cover at least c. One measure of the effectiveness of a predictor could then be taken as the long-run proportion of successes. We thus have a sequence of situations $(\theta_1, x_1), \ldots, (\theta_r, x_r)$ and the corresponding predictions $\delta(x_1), \ldots, \delta(x_r)$. If we define

$$t_\delta(\theta, x) = \begin{cases} 1 & \text{if} \quad P_f\{\delta(x) \mid \theta\} \geqslant c, \\ 0 & \text{otherwise,} \end{cases} \tag{6.12}$$

then we have a counting random variable for success. We wish the long-run proportion of success to be approximately (or at least) g, where g will not be too far from 1, so that

$$\frac{t_\delta(\theta_1, x_1) + \ldots + t_\delta(\theta_r, x_r)}{r} = g. \tag{6.13}$$

Now the model counterpart of long-run average is expectation and so we wish to choose δ such that

$$\mathrm{E}\{t_\delta(\cdot, \cdot)\} = g.$$

This means that we require

$$\int_\Theta \int_X t_\delta(\theta, x) p_e(x \mid \theta) p(\theta) \mathrm{d}\theta \mathrm{d}x = g, \tag{6.14}$$

where $p(\theta)$ is the (possibly unknown) density with which nature selects θ. Taking a frequentist approach to this we first write it in the form:

$$\int_{\Theta} p(\theta)\left\{\int_X t_\delta(\theta, x)p_e(x\,|\,\theta)\mathrm{d}x\right\}\mathrm{d}\theta = g, \tag{6.15}$$

and note that this relation will be satisfied for all $p(\theta)$ if we can arrange that the inner integral is equal to g, that is,

$$\int_X t_\delta(\theta, x)p_e(x\,|\,\theta)\mathrm{d}x = g \text{ for every } \theta \in \Theta. \tag{6.16}$$

We can express this more as a probabilistic statement in terms of P_e and P_f probability measures, for

$$\begin{aligned}\int_X t_\delta(\theta, x)p_e(x\,|\,\theta)\mathrm{d}x &= \int_{[x:\,P_f(\delta(x)|\theta)\geqslant c]} 1\cdot p_e(x\,|\,\theta)\mathrm{d}x \\ &= P_e[x: P_f\{\delta(x)\,|\,\theta\}\geqslant c\,|\,\theta].\end{aligned} \tag{6.17}$$

Thus it is reasonable to use a (c, g) predictor if a sequence of predictions is envisaged, one for each informative experiment, and if we want to guarantee that a proportion g of these regions will be reliable in the sense that they provide cover at least c.

6.3 Guaranteed coverage tolerance predictors for the binomial and Poisson cases

The easiest approach to guaranteed coverage predictors for the standard discrete distributions – the binomial and Poisson distributions – is through the theory of confidence intervals for the distribution parameter – the binomial success probability θ or the Poisson mean θ. We describe the construction of predictors of the form

$$\delta(x) = \{0, 1, \ldots, \epsilon(x)\}. \tag{6.18}$$

For any given $\theta \in \Theta$ we can determine $d_c(\theta)$, the smallest integer d satisfying the inequality

$$\sum_{y=0}^{d} p_f(y\,|\,\theta) \geqslant c. \tag{6.19}$$

Note that for both the binomial and Poisson distributions $d_c(\theta)$ is an increasing function of θ. If we knew θ then $d_c(\theta)$ could be regarded as a $(c, 1)$ tolerance predictor. It is the fact that we do not know θ that makes us lower our sights below 1 to g for the guarantee. But we can find a confidence interval for θ by standard methods. Let $u_g(x)$ be an upper confidence limit for θ at confidence level g. Then

$$P_e\{x: u_g(x) \geqslant \theta\,|\,\theta\} \geqslant g \quad (\theta \in \Theta) \tag{6.20}$$

so that

$$P_e[x : d_c\{u_g(x)\} \geqslant d_c(\theta) | \theta] \geqslant g \quad (\theta \in \Theta), \tag{6.21}$$

and if we set

$$\epsilon(x) = d_c\{u_g(x)\} \tag{6.22}$$

the interior inequality is equivalent to

$$P_f\{\delta(x) | \theta\} \geqslant c. \tag{6.23}$$

Hence δ satisfies definition 6.2 and so is a (c, g) tolerance predictor.

Binomial case. The problem remaining for special cases is the determination of the functions u_g and d_c. The basis of their determination for the binomial family lies in the relationship between the binomial distribution function and the incomplete beta function:

$$\sum_{x=0}^{d} \binom{n}{x} \theta^x (1-\theta)^{n-x} = 1 - I_\theta(d+1, n-d). \tag{6.24}$$

As in §5.5 we consider the case of a Bi(n, θ) informative experiment and a Bi(N, θ) future experiment. The upper confidence limit $u_g(x)$ for θ at confidence level g is the solution of

$$I_\theta(x+1, n-x) = g \tag{6.25}$$

for θ; alternatively Pearson and Hartley (1966) provide $u_g(x)$ directly in their table 41 for $g = 0.975, 0.995$. Moreover $d_c(\theta)$ is the minimum integer d satisfying

$$I_\theta(d+1, N-d) \leqslant 1 - c. \tag{6.26}$$

Example 6.2

A quality control problem. Recall example 1.3. For a particular batch let θ denote the probability that an item will be defective, so that we can regard e as Bi$(5, \theta)$ and a typical f (determining the number of defectives in a packet) as Bi$(25, \theta)$. If for an outcome x of e we state on each packet that the number y of defective items will be in the interval $\delta(x)$ of form (6.18) then the contract requires that

$$P_e[x : P_f\{\delta(x) | \theta\} \geqslant 0.8 | \theta] \geqslant 0.9 \tag{6.27}$$

whatever the value of θ. Comparing this with definition 6.2 we clearly require a (c, g) tolerance predictor with $c = 0.8$ and $g = 0.9$.

Consider the construction of this type of tolerance prediction corresponding to $x = 1$. First from (6.25) we have to solve

$$I_\theta(2,4) = 0.9 \tag{6.28}$$

for θ to obtain $u_g(1)$. From the Pearson (1934) tables we obtain

$$u_g(1) = 0.584. \tag{6.29}$$

Then from (6.26) we obtain $\epsilon(1)$ as the minimum integer d satisfying

$$I_{0.584}(d+1, 25-d) \leqslant 0.2. \tag{6.30}$$

From the Pearson (1934) tables we have

$$I_{0.584}(16+1, 25-16) \approx 0.22,$$

$$I_{0.584}(17+1, 25-17) \approx 0.12,$$

so that

$$\epsilon(1) = 17. \tag{6.31}$$

The complete guaranteed coverage tolerance predictor is set out in table 6.1. Note that the criterion of a (c, g) tolerance predictor is so much more demanding than a c-mean coverage tolerance predictor that much wider prediction intervals (cf. table 5.2) result.

Table 6.1 *Guaranteed coverage predictor for the quality control problem, with* $(c, g) = (0.80, 0.90)$

$\delta(x) = \{0, 1, \ldots, \epsilon(x)\}$

x	0	1	2	3	4	5
$\epsilon(x)$	11	17	21	24	25	25

Poisson case. The determination of the functions u_g and d_c for a Po$(k\theta)$ informative experiment and a Po$(K\theta)$ future experiment is similar to that for the binomial case, with the incomplete gamma function J replacing the incomplete beta function I. The basic relationship between the Poisson distribution function and the incomplete gamma function takes the form:

$$\sum_{x=0}^{d} \frac{\exp(-\theta)\theta^x}{x!} = 1 - J_\theta(d+1). \tag{6.32}$$

The upper confidence limit $u_g(x)$ for θ at confidence level g is then the solution of

$$J_{k\theta}(x+1) = g \tag{6.33}$$

for θ; for the particular cases $g = 0.95, 0.975, 0.99, 0.999$, Pearson and Hartley (1966) provide $ku_g(x)$ directly in their table 40. Similarly $d_c(\theta)$ is the minimum integer d satisfying

$$J_{K\theta}(d+1) \leqslant 1-c. \tag{6.34}$$

6.4 Guaranteed coverage tolerance predictors for the gamma case

Here as in §5.7 we are interested in finding a tolerance predictor δ, based on the information x from a $\mathrm{Ga}(k, \theta)$ experiment for the outcome of a $\mathrm{Ga}(K, \theta)$ experiment and of the form

$$\delta(x) = (0, qx). \tag{6.35}$$

The coverage is then

$$\int_0^{qx} \frac{\theta^K}{\Gamma(K)} y^{K-1} \exp(-\theta y)\, dy \tag{6.36}$$

and we have seen in §5.7 that the coverage distribution is independent of θ. In evaluating the guarantee function we can therefore conveniently set $\theta = 1$. We then have

$$\begin{aligned} g(\theta \mid \delta, c) &= P_e\left\{x : \int_0^{qx} \frac{y^{K-1}}{\Gamma(K)} \exp(-y) dy \geqslant c\right\} \\ &= P_e\{x : qx \geqslant \mathrm{Ga}(K, 1; c)\} \\ &= P_e\left\{x : x \geqslant \frac{1}{q} \mathrm{Ga}(K, 1; c)\right\}. \end{aligned} \tag{6.37}$$

We shall thus obtain the value g for this if we set $(1/q)\,\mathrm{Ga}(K, 1; c)$ equal to the $(1-g)$-quantile of the $\mathrm{Ga}(k, 1)$ distribution. Thus we have

$$\frac{1}{q} \mathrm{Ga}(K, 1; c) = \mathrm{Ga}(k, 1; 1-g)$$

and so

$$q = \frac{\mathrm{Ga}(K, 1; c)}{\mathrm{Ga}(k, 1; 1-g)}. \tag{6.38}$$

We therefore have the following result.

The predictor

$$\delta(x) = \left\{0, \frac{\mathrm{Ga}(K, 1; c)}{\mathrm{Ga}(k, 1; 1-g)} x\right\} \tag{6.39}$$

is a similar (c, g) tolerance predictor.

The corresponding (c, g) tolerance predictor δ of the form

$$\delta(x) = (qx, \infty) \tag{6.40}$$

can easily be shown to have

$$q = \frac{\mathrm{Ga}(K, 1; 1-c)}{\mathrm{Ga}(k, 1; g)}. \tag{6.41}$$

Many (c, g) finite interval predictors can be obtained. For

$$\delta(x) = [q_1 x, q_2 x] \tag{6.42}$$

we have a (c, g) similar guaranteed coverage tolerance predictor if we choose

$$q_1 = \frac{\mathrm{Ga}(K, 1; c_1)}{\mathrm{Ga}(k, 1; g_2)}, \quad q_2 = \frac{\mathrm{Ga}(K, 1; c_2)}{\mathrm{Ga}(k, 1; g_1)}, \tag{6.43}$$

where $c_2 - c_1 = c, g_2 - g_1 = g$. The choice of specific $c_1, c_2, g_1 . g_2$ will depend on what other feature, such as shortness of interval, we require of the predictor.

6.5 Guaranteed coverage tolerance predictors for the normal case

We suppose that the future experiment and the informative experiment are as described in the corresponding section §5.8 on mean coverage tolerance predictors.

One-sided tolerance predictors. Consider a predictor δ, based on the information (m, v) as specified in §5.8, and of the form

$$\delta(m, v) = (-\infty, m + qv^{1/2}). \tag{6.44}$$

The coverage is then $\Phi\{(K\tau)^{1/2}(m + qv^{1/2} - \mu)\}$ and we have seen in our considerations of mean coverage tolerance predictors that the coverage distribution is independent of $\theta = (\mu, \tau)$, so that we can consider the case $\theta = (0, 1)$ in evaluating the guarantee function associated with δ. We then have

$$\begin{aligned} g(\theta \mid \delta, c) &= P_{ef}[(m, v) \colon \Phi\{K^{1/2}(m + qv^{1/2})\} \geqslant c] \\ &= P_{ef}\{(m, v) \colon K^{1/2}(m + qv^{1/2}) \geqslant \Phi^{-1}(c)\} \\ &= P_{ef}\left[(m, v) \colon \left\{-mk^{1/2} + \left(\frac{k}{K}\right)^{1/2} \Phi^{-1}(c)\right\}\left(\frac{\nu}{v}\right)^{1/2} \leqslant (k\nu)^{1/2} q\right]. \end{aligned} \tag{6.45}$$

Since m and v can be taken to be independently distributed as $\mathrm{No}(0, k)$ and $\mathrm{Ch}(\nu, 1)$ respectively the statistic on the left side of the inequality in (6.45) has the non-central t-distribution $\mathrm{t}\{\nu, (k/K)^{1/2} \Phi^{-1}(c)\}$. (See Resnikoff and Lieberman, 1957, and Resnikoff, 1962.) In order to obtain a similar (c, g) predictor, that is with

$$g(\theta \mid \delta, c) = g \text{ for every } \theta \in \Theta,$$

we simply set

$$q(k\nu)^{1/2} = \mathrm{t}\left\{\nu, \left(\frac{k}{K}\right)^{1/2} \Phi^{-1}(c); g\right\}.$$

We thus have the following results.

The predictor

$$\delta(m, v) = \left[-\infty, m + \left(\frac{v}{k\nu}\right)^{1/2} \mathrm{t}\left\{\nu, \left(\frac{k}{K}\right)^{1/2} \Phi^{-1}(c); g\right\}\right] \tag{6.46}$$

is a similar (c, g) tolerance predictor.

Similarly there can be derived the following result for the other kind of one-sided region.

The predictor

$$\delta(m, v) = \left[m - \left(\frac{v}{k\nu}\right)^{1/2} \mathrm{t}\left\{\nu, \left(\frac{k}{K}\right)^{1/2} \Phi^{-1}(c); g\right\}, \infty\right] \tag{6.47}$$

is a similar (c, g) tolerance predictor.

Symmetric two-sided tolerance predictors. For a predictor

$$\delta(m, v) = (m - qv^{1/2}, m + qv^{1/2}) \tag{6.48}$$

the coverage is

$$\Phi\{(K\tau)^{1/2}(m + qv^{1/2} - \mu)\} - \Phi\{(K\tau)^{1/2}(m - qv^{1/2} - \mu)\}.$$

Here again the guarantee function is constant for all $\theta = (\mu, \tau)$ and so the problem is essentially that of evaluating

$$P_e[(m, v): \Phi\{K^{1/2}(m + qv^{1/2})\} - \Phi\{K^{1/2}(m - qv^{1/2})\} \geqslant c], \tag{6.49}$$

again with m and v independently distributed as $\mathrm{No}(0, k)$ and $\mathrm{Ch}(\nu, 1)$. The problem is thus the computational one of evaluating the double integral of $p(m)p(v)$ over the region of integration R shown in fig. 6.2. The shape is determined by the following factors concerning the coverage

$$\Phi\{K^{1/2}(m + qv^{1/2})\} - \Phi\{K^{1/2}(m - qv^{1/2})\}.$$

(i) The coverage is a strictly increasing function of v.
(ii) The coverage is a symmetric function in m.
(iii) For fixed m, $\Phi\{K^{1/2}(m + qv^{1/2})\} - \Phi\{K^{1/2}(m - qv^{1/2})\} = c$ has a unique solution in v.

The solution of the computational problem here has had a long and interesting history (Wald and Wolfowitz, 1946; Wallis, 1951; Lieberman, 1957; Weissberg and Beatty, 1960; Ellison, 1964). The most recent and very satisfactory method for the computation of q for given k and ν is given by Howe (1969) and uses only tables of normal and chi-squared quantiles. Fig. 6.3 shows the flow chart for the computation of q and for determining the order of the approximation in terms of the closeness of the guarantee attained to the guarantee aimed at. For example, if $\nu \leqslant \nu_g$ and we use the q shown, then

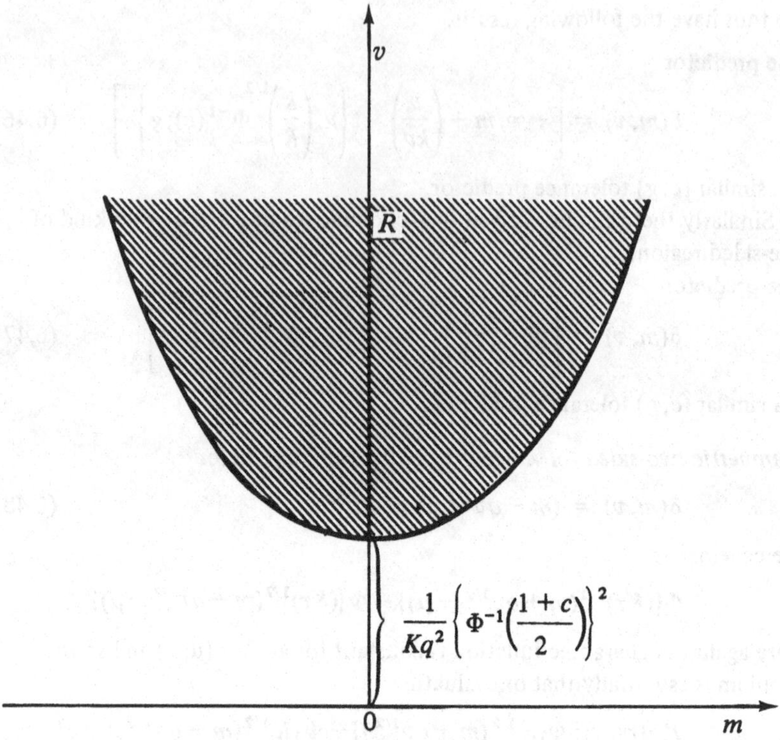

Fig. 6.2 Region of integration R in (6.49) for which $\Phi\{K^{1/2}(m+qv^{1/2})\}-\Phi\{K^{1/2}(m-qv^{1/2})\}\geqslant c$.

$$P_e[(m,v)\colon \Phi\{K^{1/2}(m+qv^{1/2})\}-\Phi\{K^{1/2}(m-qv^{1/2})\}\geqslant c]$$
$$= g + O\left\{\max\left(\frac{K^2v^{1/2}}{k^2},\frac{K^3v^{3/2}}{k^3}\right)\right\}. \quad (6.50)$$

Howe (1969) also shows numerically that the order of approximation is often very much better than that ascribed by this general result, especially for cases where ν and k^2 are of the same order of magnitude. The route along the flow chart depends on the relative magnitudes of the effective sample size k associated with the estimate of the mean μ and the number ν of degrees of freedom associated with the estimate of variance.

Example 6.3

Contract for design components. The distributions of the characteristics of components from a simultaneous production run on each of four different machines are known from past experience to be well approximated by normal

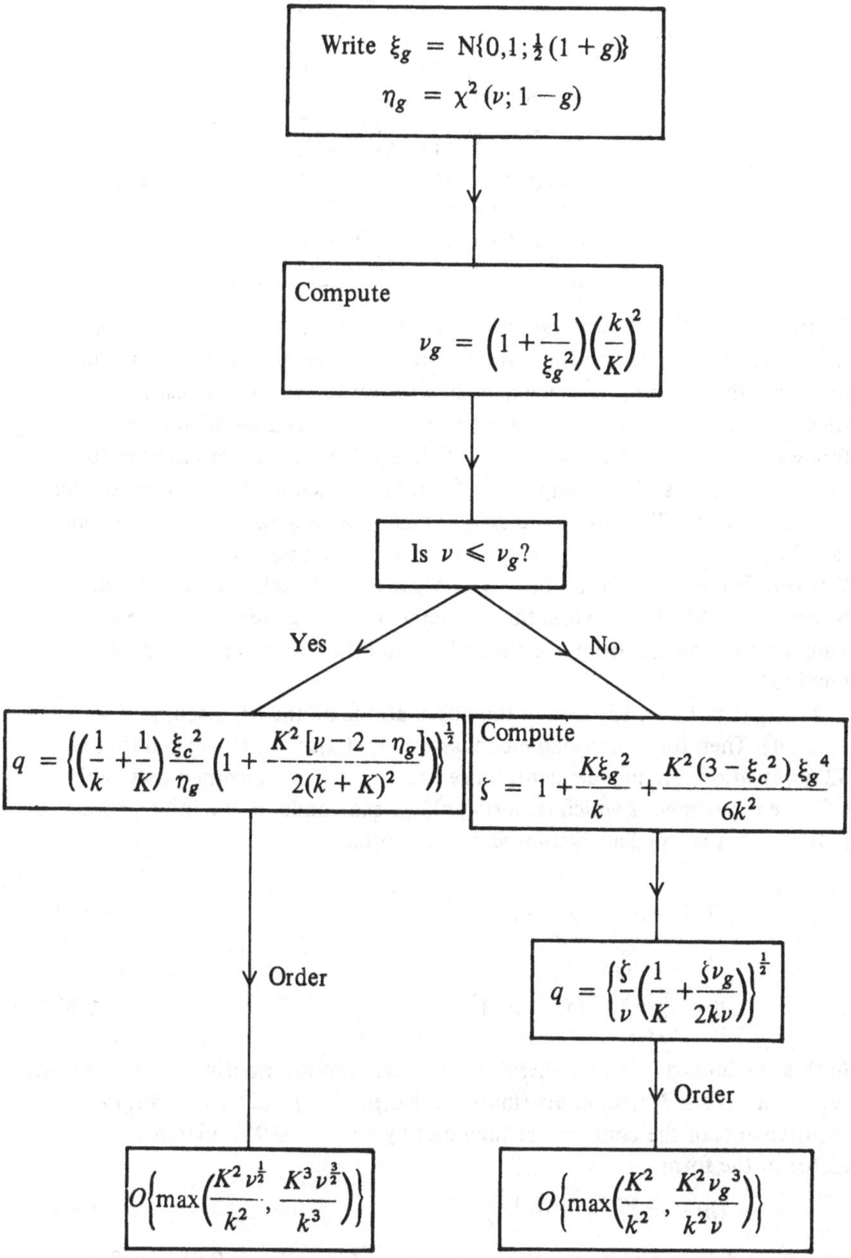

Fig. 6.3 Flow chart for computation of two-sided normal tolerance limits.

Table 6.2 *Characteristics of the 52 sampled components*

Machine	Characteristics
1	51.3, 51.2, 48.4, 46.1, 51.6, 51.4, 51.7, 48.1, 51.6, 46.6, 49.2, 46.5, 48.0
2	56.2, 57.3, 54.5, 53.6, 51.7, 55.5, 54.1, 53.0, 52.1, 51.3, 54.6, 55.1, 53.2
3	59.0, 59.7, 58.2, 56.7, 59.6, 60.2, 61.5, 62.5, 61.3, 58.5, 60.6, 61.3, 57.7
4	49.1, 49.6, 44.5, 45.9, 46.0, 48.2, 49.6, 52.5, 48.8, 45.6, 47.2, 45.6, 47.7

distributions with the same variance. The distribution means, however, vary from machine to machine and, even for the same machine, from run to run; moreover the common variance may differ from run to run. A designer contracts to accept the complete output from the first machine subject to the following conditions. The producer will determine for each production run the characteristics of 13 components from each machine. On the basis of this information he will quote to the designer an interval which purports to contain 95 per cent of the component characteristics supplied to the designer from that run. It is also agreed that at least 95 per cent of such quotations should be correct. Table 6.2 provides the characteristics associated with the 52 sampled components from one run. What quotation should the producer provide?

Let x_{ir} $(r = 1, \ldots, 13)$ denote the observations on the ith machine $(i = 1, \ldots, 4)$. Then the underlying model asserts that x_{ir} is $\mathrm{No}(\mu_i, \tau)$ and that all 52 observations are independent. Since the contract is concerned only with a future experiment f which is $\mathrm{No}(\mu_1, \tau)$, we can condense the information in the x_{ir} in the usual analysis-of-variance approach to

$$m = x_{1\cdot} = \sum_{r=1}^{13} x_{1r}/13, \tag{6.51}$$

$$v = \sum_{i=1}^{4} \sum_{r=1}^{13} (x_{ir} - x_{i\cdot})^2. \tag{6.52}$$

In this condensation we are therefore envisaging an informative experiment of type 5 in table 2.3, with m distributed as $\mathrm{No}(\mu_1, 13\tau)$ and v as $\mathrm{Ch}(48, \tau)$. The requirements of the contract are then met by a $(0.95, 0.95)$ tolerance predictor of the form

$$(m - qv^{1/2}, m + qv^{1/2}), \tag{6.53}$$

where q is determined by the flow chart of fig. 6.3 with $c = 0.95, g = 0.95, k = 13, \nu = 48, K = 1$.

Since

$$\nu_g = \left(1 + \frac{1}{(1.96)^2}\right)(13)^2 = 213 \tag{6.54}$$

we have $\nu \leqslant \nu_g$ and so the computations follow the left-hand branch, leading to

$$q = 0.359. \tag{6.55}$$

From the data of table 6.2,

$$m = 49.36, \ v = 188.96 \tag{6.56}$$

so that the appropriate quotation to the designer is the interval

$$(44.4, 54.3). \tag{6.57}$$

The order of the approximation computed from the flow diagram is here

$$\max\left(\frac{(48)^{1/2}}{(13)^2}, \frac{(48)^{3/2}}{(13)^3}\right) = 0.15, \tag{6.58}$$

which appears remarkably poor. Howe shows, however, that for this particular ν, k and K the actual guarantee is 0.9526 compared with the specified target of 0.95.

6.6 Guaranteed coverage tolerance predictors for the multinormal case

It is clear that the already non-trivial computational problem of constructing guaranteed coverage tolerance predictors for the univariate case of §6.5 becomes even more complicated for the multinormal case. The simplest form of predictor to consider is again the elliptical type of form (5.56). A successful computational technique, due to Guttman (1970), is

(i) to obtain, for given q, approximations to the mean and variance of the coverage distribution,
(ii) to 'fit' a beta distribution, say $\mathrm{Be}\{r(q), s(q)\}$, with the same mean and variance,
(iii) to 'adjust' q until the fitted beta distribution has its g-quantile equal to c, namely

$$I_c\{r(q), s(q)\} = g. \tag{6.59}$$

For details and tables of q for $d = 2, 3, 4$ and $c, g = 0.75, 0.90, 0.95, 0.99$, see Guttman (1970).

6.7 Guaranteed coverage tolerance predictors for the two-parameter exponential case

We now consider the construction of (c, g) tolerance predictors when the basic situation is of two-parameter exponential type as described in §5.10.

To obtain an interval predictor of the form

$$(m + qv, \infty) \tag{6.60}$$

which gives coverage at least c with guarantee g we require that

$$P_e\left\{(m, v): \int_{m+qv}^{\infty} p(y)\mathrm{d}y \geqslant c\right\} = g \tag{6.61}$$

or

$$P_e\left\{(m, v): m + qv \leqslant -\frac{1}{K} \log c\right\} = g. \tag{6.62}$$

(Here as in §5.10 the parameters (μ, τ) can be set equal to $(0, 1)$.) The awkward computational aspect here is that the form of the left hand side depends crucially on the relative magnitudes of c and g. This dependence can be clearly seen in table 6.3 which gives the guarantee g corresponding to given c and q. Note how the form of this measure depends on the value of q. The results are easily established when the double integral is expressed as a repeated integral.

Table 6.3 *Guarantee associated with* $(m + qv, \infty)$ *for two-parameter exponential distribution*

Range of q	Guarantee
$q \leqslant 0$	$1 - \dfrac{c^{k/K}}{(1 - kq)^{\nu}}$
$0 < q < \dfrac{1}{K}$	$J_{-[1/(Kq)] \log c}(\nu) - \dfrac{c^{k/K}}{(1 - kq)^{\nu}} J_{-[(1-kq)/(Kq)] \log c}(\nu)$
	or $\displaystyle\int_0^{-\frac{1}{K}\log c} J_{(-\log c - Km)/(Kq)}(\nu)\, p(m)\,\mathrm{d}m$
$q = \dfrac{1}{k}$	$J_{-(k/K) \log c}(\nu + 1)$
$q > \dfrac{1}{k}$	$J_{-[1/(Kq)] \log c}(\nu) - c^{k/K} \displaystyle\int_0^{-\frac{1}{Kq}\log c} \exp(kqv)\, p(v)\,\mathrm{d}v$

The computational problem is, of course, to determine q for given c, g, k, ν and K, but table 6.3 shows that the form of q will depend on the relative magnitudes of c, g, k, ν and K. The table can be converted into a flow diagram (fig. 6.4) to determine which form of table 6.3 should be used. From the first form the guarantee of the interval predictor (m, ∞) is $1 - c^{k/K}$. Hence, if

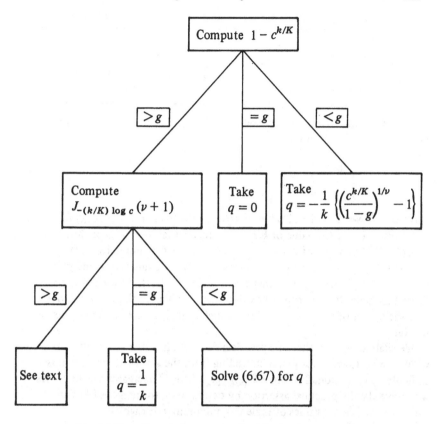

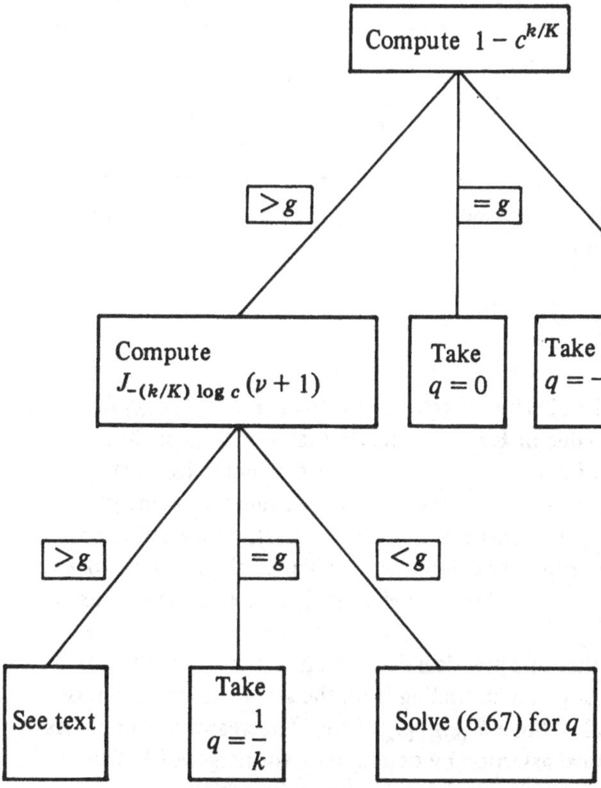

Fig. 6.4 Flow diagram for computation of a (c, g) guaranteed coverage tolerance interval for the two-parameter exponential distribution.

$1-c^{k/K}=g$, we take $q=0$. If $1-c^{k/K}<g$ we have not achieved the desired guarantee and so will have to widen the interval (m, ∞) by taking a negative q. Solving

$$1-\frac{c^{k/K}}{(1-kq)^{\nu}}=g \tag{6.63}$$

for q then provides the stated formula for q. If $1-c^{k/K}>g$ then we shall have to shorten the interval (m, ∞), thus taking a positive q. Note that (m, ∞) has acted as a criterion interval at this first node of the flow diagram. At the second node a similar type of argument applies, the criterion interval $(m+(1/k)v, \infty)$ of guarantee $J_{-(k/K)\log c}(\nu+1)$ providing the basis for the routing of the computations.

At first sight some of the computations seem rather laborious but in what follows we shall show that the worst is most unlikely to arise in practice, and we shall give a simple desk computer method for the others.

Table 6.4 *Maximum k/K for which second node of fig. 6.4 is not reached*

	c		
g	0.9	0.95	0.99
0.9	21	44	229
0.95	28	58	298
0.99	43	89	458

First we shall arrive at the second node (and so the awkward-looking computations) if and only if

$$\frac{k}{K} > \frac{\log(1-g)}{\log c}. \tag{6.64}$$

Table 6.4 shows the critical value of k/K for practical values of (c, g) in the following sense: if the value of k/K is no greater than the number shown then, regardless of the value of ν, we have $q \leqslant 0$ and use the simple form. (We shall assume throughout that interest is in reasonably high values of c and g, say $c \geqslant 0.9, g \geqslant 0.9$; the selection made is intended to illustrate the concepts.) From this table the experimenter will be aware when he chooses k relative to K at the outset of his experiment just which branch of the computations he will face.

We shall now show that in a practical situation it is most unlikely that we would travel down the $>g$ branch leading from the second node (the most difficult computationally). Since $J_{-(k/K)\log c}(\nu+1)$ decreases as ν increases we can verify the practical assertion by determining for ranges of k/K in excess of the critical values of table 6.4, minimum ν for which

$$J_{-(k/K)\log c}(\nu+1) < g. \tag{6.65}$$

Table 6.5 gives these minima for $g = 0.9$. For values of g higher than 0.9 the critical value of ν is less than the corresponding value for $g = 0.9$. To place this in a life-testing context we would expect an experimenter wishing to obtain reasonably satisfactory insight into the properties of his items to see a life-testing experiment through to a value of ν at least that shown in the table. For example, at the (0.9, 0.9) level it seems most unlikely that in testing 50 items he would be happy with his experiment if he had not seen through to failure at least $r = \nu + 1 = 3$ items for the case $K = 1$. We have thus taken the view that it is best to concentrate our energies in the investigation of the computation arising from the $<g$ branch leading from the second node. The computations here are within the scope of speedy computation on a desk computer. (Along the $>g$ branch it seems that numerical integration is necessary, which complicates the computation, although it is well within the scope of a moderate automatic computer.)

Table 6.5 *Minimum ν for which guarantee of* $(m + (1/k)\nu, \infty)$ *is less than* $g = 0.9$

c = 0.9		c = 0.95		c = 0.99	
k/K	ν	k/K	ν	k/K	ν
25	1	50	1	300	1
50	2	100	2	400	2
100	7	200	6	500	2
200	15	300	10	600	3

To complete this investigation we have therefore to consider, for values of ν exceeding those of table 6.5, or more precisely for configurations of (c, g, k, K, ν) satisfying

$$J_{-(k/K)\log c}(\nu + 1) < g < 1 - c^{k/K}, \tag{6.66}$$

the solution of the equation

$$G(q) = g, \tag{6.67}$$

where

$$\begin{aligned} G(q) &= G_{k,K,\nu,c}(q) \quad \left(0 < q < \frac{1}{k}\right) \\ &= J_{-[1/(Kq)]\log c}(\nu) - \frac{c^{k/K}}{(1-kq)^{\nu}} J_{-[(1-kq)/(Kq)]\log c}(\nu). \end{aligned} \tag{6.68}$$

Now

$$\begin{aligned} G'(q) &= \frac{d}{dq} \int_0^{-\frac{1}{K}\log c} \left\{ \int_0^{\frac{-\log c - Km}{Kq}} p(v)dv \right\} p(m)dm \\ &= -\frac{1}{\Gamma(\nu)} \int_0^{-\frac{1}{K}\log c} \frac{(-\log c - Km)}{Kq^2} \left(\frac{-\log c - Km}{Kq}\right)^{\nu-1} \\ &\quad \times \exp\left\{-\left(\frac{-\log c - Km}{Kq}\right)\right\} k \exp(-km)dm \\ &= -\frac{\nu k c^{k/K}}{(1-kq)^{\nu+1}} J_{-[(1-kq)/(Kq)]\log c}(\nu + 1). \end{aligned} \tag{6.69}$$

Since $G'(q) < 0$ $(0 < q < 1/n)$ the function $G(q)$ steadily decreases from $1 - c^{k/K}$ through g to $J_{-(k/K)\log c}(\nu + 1)$.

We have expressed $G(q)$ and $G'(q)$ very simply in terms of incomplete gamma ratio functions J, and so, with a suitable table of these functions (Pearson, 1922), or with a suitable subroutine, for example, using

$$J_x(\nu) = 1 - \exp(-x)\left(1 + \frac{x}{1!} + \ldots + \frac{x^{\nu-1}}{(\nu-1)!}\right), \tag{6.70}$$

for their computation, q can very quickly be determined by the Newton–Raphson iterative procedure for the solution of (6.67), namely

$$q_r = q_{r-1} + \frac{g - G(q_{r-1})}{G'(q_{r-1})}. \tag{6.71}$$

For the range of c and g already described, for the range of k/K in table 6.4, and for the complete range of ν exceeding the values quoted in table 6.5, we have explored numerically this computational procedure. As a consequence of this investigation we recommend as initial value for the iterative procedure:

$$q_0 = \frac{1}{k} - \frac{1}{k}\left(\frac{c^{k/K}}{1-g}\right)^{1/\nu}.$$

This clearly lies in the correct interval $0 < q < 1/k$ and is in fact the solution that we would obtain by setting both J values equal to 1 in (6.68).

History

The concept of a guaranteed coverage tolerance predictor was first considered by Wald and Wolfowitz (1946) who also obtained a solution to the computational problem for the finite interval normal case. As already mentioned in §6.5 the development of this particular problem was continued through the work of Wallis (1951), who showed its extension from the case $\nu = k - 1$ to the regression situation; Lieberman (1957) and Weissberg and Beatty (1960), who provided tables for q; Ellison (1964) and Howe (1969), who allow the direct computation through the flow chart of fig. 6.3. As already mentioned Guttman (1970) provides a treatment of the multinormal case, and also of the Ex(θ) case. For other distributions the development is less clear. For tables and applications for the binomial, Poisson and gamma cases see Aitchison (1963) and Aitchison and Sculthorpe (1964). As far as we can determine the treatment of the two-parameter exponential distribution given in §6.7 is new. See also Aitchison (1964) for a discussion of the relationship of guaranteed coverage predictors to a Bayesian form of (c, g) prediction, not discussed in this book.

Problems

6.1 For the case where the informative experiment e and the future experiment f are both described by Ge(θ) random variables, derive guaranteed coverage tolerance predictors of the form

(i) $\delta(x) = \{0, 1, \ldots, \epsilon(x)\}$,

(ii) $\delta(x) = \{\epsilon(x), \epsilon(x)+1, \ldots\}$.

6.2 (i) For the binomial case where e, f are described by $\mathrm{Bi}(n, \theta)$, $\mathrm{Bi}(N, \theta)$ random variables, derive a guaranteed coverage tolerance predictor of the form

$$\delta(x) = \{\epsilon(x), \epsilon(x) + 1, \ldots, N\}.$$

(ii) For the Poisson case where e, f are described by $\mathrm{Po}(k\theta)$, $\mathrm{Po}(K\theta)$ random variables, derive a guaranteed coverage tolerance predictor of the form

$$\delta(x) = \{\epsilon(x), \epsilon(x) + 1, \ldots\}.$$

6.3 Derive (c, g) guaranteed coverage tolerance predictors of the form $(k + qv, \infty)$ for the Pareto distribution described in problem 5.2.

6.4 An ornithologist wishes to predict the numbers of eggs laid by colonies of birds in different locations. Suppose that the numbers laid by different females in a year are described by independent $\mathrm{Po}(\theta)$ random variables, where θ may vary with location. The ornithologist obtains information by selecting a random sample of 20 nests in a colony. Suppose these yield a total of 43 eggs. Provide a statement concerning the number of eggs laid by a bird in the colony which will be correct for at least 90 per cent of the birds in the colony and which will be true for 95 per cent of colonies.

6.5 Complete the analysis of problem 1.3.

6.6 Close study of ten patients discharged from hospital after treatment for a chronic disorder has been undertaken. The times to relapse of the first six to relapse out of the ten patients were 45, 52, 67, 84, 108, 135 days, the remaining four patients showing no sign of relapse at the 135th day since discharge. The clinic is now trying to formulate a reasonable policy for the recall time of patients which, for administrative purposes, must be the same for all patients. It wants to be reasonably certain (say 95 per cent) that 95 per cent of patients are recalled for treatment before relapse. On the assumption that times to relapse have a two-parameter exponential distribution what policy do you recommend?

6.7 Reconsider problem 5.5. Suppose now that the light bulbs are produced in larger batches than 50. Before marketing the manufacturer wishes to make a statement to the effect that at least 95 per cent of the bulbs in a batch will have lifetimes greater than a given value, and he wishes to be correct in his statements for 95 per cent of all batches. Obtain the relevant guaranteed coverage tolerance predictor for the data provided.

Find the limiting tolerance predictor of mean coverage 0.95 as the size of the batch increases from 50.

6.8 A shoe manufacturer assumes that the distribution of foot length and foot breadth in his women customers is bivariate normal. The relevant information from a sample study of 200 women customers is shown below. For shoe size 6 designed to meet the requirements of his women customers of foot length 26 cm the manufacturer wants to cater in breadth sizes for 95 per cent of these customers. For what range of foot breadth should he design to be 90 per cent certain of meeting this requirement?

	Foot length (cm)	Foot breadth (cm)
Sample mean	24.3	8.5
Sample standard deviation	3.4	1.6

Sample correlation coefficient between two measurements: 0.62.

7
Other approaches to prediction

7.1 Introduction

In this chapter we discuss some further aspects concerning decisive and informative prediction. We recall that the decisive prediction approach can be used when knowledge of the predictive distribution $p(y|x)$ is available together with a utility function $U(a, y)$. When no utility specification is available one has to use an informative prediction approach and obtain either a most plausible Bayesian predictor or a frequentist tolerance predictor.

7.2 Linear utility structure

There is an interesting tie up between decisive prediction and Bayesian informative prediction for the case where we can assume a linear utility structure. For example suppose that we take the linear loss structure (3.7), that is

$$U(a, y) = \begin{cases} -\xi(a-y) & (y<a), \\ -\eta(y-a) & (y \geqslant a), \end{cases} \tag{7.1}$$

for the prediction of a one-sided interval $(-\infty, a)$. From equation (3.8) we have that the optimal a^* is given by the solution of

$$\int_{-\infty}^{a^*} p(y|x)\mathrm{d}y = \frac{\eta}{\xi+\eta}. \tag{7.2}$$

This interval obviously corresponds to an informative Bayesian prediction interval (not necessarily the most plausible) of coverage $\kappa = \eta/(\xi + \eta)$. The provision of this alternative view of Bayesian cover intervals may make them more attractive to some users in that they feel that it is easier to assess the relative cost factor η/ξ than the more nebulous cover κ. For example a statistician who uses a 95 per cent cover interval is behaving in approximately the same way as a statistician who regards the proportional loss caused by outcomes above the limit to be 19 times more serious than that caused by outcomes inside the interval. Table 7.1 illustrates the relationship between η/ξ and κ for several cases.

We may investigate a similar relationship for a linear loss structure of the

Table 7.1 *Bayesian cover κ corresponding to relative cost factor η/ξ for utility function (7.1)*

η/ξ	κ
0.5	0.33
1	0.50
3	0.75
9	0.90
19	0.95
99	0.99

form (3.28), that is,

$$U(a_1, a_2, y) = \begin{cases} -\xi(a_1 - y) - (a_2 - a_1) & (y < a_1), \\ -(a_2 - a_1) & (a_1 \leqslant y \leqslant a_2), \\ -\eta(y - a_2) - (a_2 - a_1) & (y > a_2), \end{cases} \quad (7.3)$$

for the prediction of a two-sided interval (a_1, a_2). From equations (3.29) we have that the optimal $(a_1{}^*, a_2{}^*)$ are given by the solutions of

$$\int_{-\infty}^{a_1{}^*} p(y|x)dy = 1/\xi, \int_{a_2{}^*}^{\infty} p(y|x)dy = 1/\eta, \quad (7.4)$$

provided $1/\xi + 1/\eta \leqslant 1$. This interval has Bayesian coverage $\kappa = 1 - 1/\xi - 1/\eta$. Thus, for example, if the differential losses associated with being outside the interval are 40 times greater than the cost per unit length of interval used, then the interval corresponds to a 95 per cent cover interval. Table 7.2 gives the Bayesian cover associated with the optimal intervals obtained for several (ξ, η) values.

Table 7.2 *Bayesian cover κ corresponding to (ξ, η) for utility function (7.3)*

ξ	η	κ	ξ	η	κ
2	2	0	2	20	0.45
4	4	0.50	20	4	0.70
20	20	0.90	20	40	0.925
40	40	0.95	20	100	0.94
200	200	0.99	200	100	0.985

7.3 Frequentist decision theory

If we are unwilling to assign a prior plausibility function $p(\theta)$ on Θ, the predictive density function $p(y|x)$ cannot be determined, and so the Bayesian decisive prediction approach of chapter 3 is no longer available. The corresponding frequentist utility approach lies along a path beset with difficulties. Although the utility $U(a, y)$ is an appealing one to the practical man it has some considerable conceptual difficulties for the frequentist. If a prediction is required for only one performance of f it is natural, in accord with frequentist decision theory, to introduce a predictor δ and take the expectation of $U(\delta(x), y)$ with respect to the informative density $p(x|\theta)$. The resulting expectation depends on (θ, y), the unknown state of nature; the presence of both θ and y makes the usual awkward feature of frequentist theory, namely the difficulty of finding pivotal statistics, even more embarrassing. The dependence of y on θ through $p(y|\theta)$ is waiting to be used but there is no obvious frequentist way to introduce it.

If a series of replicates of f is to be conducted and the prediction region to be used in each replicate, the frequentist could then obtain the induced utility $V(\delta(x), \theta)$ from

$$V(\delta(x), \theta) = \int_Y U(\delta(x), y)\, p(y|\theta)\mathrm{d}y. \tag{7.5}$$

He would then proceed in the usual way by basing his considerations on

$$G(\delta, \theta) = \int_X V\{\delta(x), \theta\}\, p(x|\theta)\mathrm{d}x \tag{7.6}$$

and searching to find an optimal δ *for all* θ – if one exists. The method implicitly requires that utilities are additive over replicates and in fact $V(\delta(x), \theta)$ corresponds to the average utility per replicate, thus providing a measure of the effectiveness of the prediction region. However this approach can lead to an inconsistency in interpretation (see Aitchison and Sculthorpe, 1965) and there is little doubt that the frequentist is on the safer logical ground – though farther removed from practical considerations – when he confines himself to utility functions of the form $V(a, \theta)$ without recourse to the more practical $U(a, y)$.

7.4 Frequentist linear utility theory

As an example where it is in fact possible to maximise $G(\delta, \theta)$ for all θ we consider the often amenable normal case with a linear utility function. We take $p(y|\theta)$ to be a $\mathrm{No}(\mu, K\tau)$ density and suppose that the informative experiment e provides us with a sufficient statistic (m, v) for (μ, τ) – so that $x = (m, v)$ – and that m and v are independent with $\mathrm{No}(\mu, k\tau)$ and $\mathrm{Ch}(\nu, \tau)$ distributions. In such circumstances we saw in chapter 5 that a tolerance interval of mean cover c of the form $(-\infty, m + qv^{1/2})$ has

$$q = \left\{\frac{1}{\nu}\left(\frac{1}{k} + \frac{1}{K}\right)\right\}^{1/2} \mathrm{t}(\nu; c). \tag{7.7}$$

We now find the frequentist decisive predictor of the form $\delta(x) = (-\infty, a)$ when we have available a utility function

$$U(\delta(x), y) = \begin{cases} -\xi(a-y) & (y<a), \\ -\eta(y-a) & (y \geqslant a). \end{cases} \tag{7.8}$$

Just as in mean cover analysis it seems sensible to consider limits of the form $a = \delta(m, v) = m + qv^{1/2}$. We attempt to find a q, if any, which maximises $G(q, \theta)$ for all θ; note that we write $G(q, \theta)$ for $G(\delta, \theta)$ since q completely specifies δ. It follows that

$$G(q, \theta) = \int_{-\infty}^{\infty} \int_0^{\infty} V(m + qv^{1/2}, \theta)\, p(m, v|\theta) dm\, dv, \tag{7.9}$$

where

$$V(m + qv^{1/2}, \theta) = \eta(m + qv^{1/2} - \mu) - (\xi + \eta) \times \int_{-\infty}^{m+qv^{1/2}} (m + qv^{1/2} - y)\, p(y\,|\theta) dy. \tag{7.10}$$

It is fairly easy to establish that a maximising value of q necessarily occurs where the derivative of $G(q, \theta)$ with respect to q is zero. The derivative equation is obtained as

$$\int_{-\infty}^{\infty} \int_0^{\infty} \{\eta - (\xi + \eta) P_f(m + qv^{1/2}|\theta)\} v^{1/2} p(m, v|\theta) dm dv = 0, \tag{7.11}$$

where $P_f(m + qv^{1/2}|\theta)$ is an abbreviation for $P_f\{(-\infty, m + qv^{1/2})|\theta\}$. Now

$$P_f(m + qv^{1/2}|\theta) = \Phi\{(K\tau)^{1/2}(m - \mu + qv^{1/2})\}.$$

If we introduce the change of variables

$$M = (k\tau)^{1/2}(m - \mu), \quad V = \tau v,$$

we reduce the left-hand side of (7.11), after the cancellation of a factor to

$$\int_{-\infty}^{\infty} \int_0^{\infty} \left[\eta - (\xi + \eta)\Phi\left\{K^{1/2}\left(\frac{M}{k^{1/2}} + qV^{1/2}\right)\right\}\right] p(M)\, p(V) dM dV, \tag{7.12}$$

where $p(M)$ and $p(V)$ are No(0, 1) = N (0, 1) and Ch$(\nu + 1, 1) = \chi^2(\nu + 1)$ densities respectively. Note that we use $p(V)$ here to denote not the density function of $V = \tau v$ which is $\chi^2(\nu)$ but the naturally arising factor in V in the integrand which turns out to be a $\chi^2(\nu + 1)$ density function. In considering (7.12) we can therefore treat M and V as independent with the specified densities and if we introduce another N(0, 1) variable W, say, independent of (M, V), we see that the derivative equation (7.11) yields

$$Pr\left\{W < K^{1/2}\left(\frac{M}{k^{1/2}} + qV^{1/2}\right)\right\} = \frac{\eta}{\xi + \eta}.$$

It follows immediately, since

$$\frac{(W/K^{1/2} - M/k^{1/2})(\nu + 1)^{1/2}}{\left(\dfrac{1}{k} + \dfrac{1}{K}\right)^{1/2} V^{1/2}}$$

is distributed as $\mathrm{t}(\nu + 1)$, that

$$q = \left\{\left(\frac{1}{k} + \frac{1}{K}\right) \Big/ (\nu + 1)\right\}^{1/2} \mathrm{t}\{\nu + 1; \eta/(\xi + \eta)\}. \qquad (7.13)$$

Thus the upper normal linear utility frequentist limit is $m + qv^{1/2}$, where q is given by (7.13).

There is here no relation between c and η/ξ (independent of ν) which leads to equality of expressions (7.7) and (7.13); the complete equivalence of the informative prediction and linear utility intervals, discussed in §7.2 for the Bayesian approach, is thus absent. When, however, the statistic v is based on a large sample so that ν is appreciable, we see from (7.7) and (7.13) that the two intervals are for all practical purposes the same if $c = \eta/(\xi + \eta)$. Since the direct interpretation of c is not without difficulty (see Aitchison and Sculthorpe, 1965, p. 477) the new interpretation arising from this correspondence provides a useful alternative view of mean cover intervals. An interval with mean cover c is for appreciable ν the same as a linear utility interval with η/ξ factor equal to $c/(1 - c)$. Table 7.1 can again be used to illustrate the relationship if we substitute expected cover c for Bayesian cover κ. For example, a statistician who uses a 95 per cent expected cover interval is behaving in approximately the same way as a statistician who regards the proportional loss caused by outcomes above the limit to be 19 times more serious than that caused by outcomes inside the interval.

While we have developed the theory for intervals of type $(-\infty, a)$ it is clear that a similar development is possible for intervals of type (a, ∞), and that the comments on correspondence between mean cover and linear utility intervals remain unchanged.

7.5 Prior probabilities on the sample space

In this section we briefly investigate an alternative approach to Bayesian prediction problems which in some cases may be more palatable to the practical man.

The possible probabilistic descriptions of the experiment f form a class of density functions $\{p(\cdot|\theta) : \theta \in \Theta\}$ on the sample space Y and indexed by parameters in the parameter space Θ. We have adopted the common (Bayesian) practice of developing the analysis by the use of a prior density function $p(\theta)$ on Θ. It does not seem to be generally realised, although it is implicit in the concepts of conditional probability and was discussed by Raiffa and Schlaifer (1961, chapter 1), that there is automatically induced a prior density $p(y)$ on Y through the relationship

$$p(y) = \int_{\Theta} p(y|\theta)\, p(\theta) d\theta. \tag{7.14}$$

(If it is impossible to perform any informative experiment then the only source of information is the prior plausibility function $p(\theta)$. We would then use $p(y)$ as the predictive density function.)

The notion of the induced prior $p(y)$ suggests an alternative Bayesian approach with a different starting point. A practical man may find it easier to specify a prior $p(y)$ than to specify a prior $p(\theta)$. He may do this in an approximate way by constructing a histogram for the interpretation of his thoughts on the likely values of y. To a Bayesian the density function of importance is $p(y|x)$. The following question now arises: given $p(y)$, can one determine $p(y|x)$?

The solution of such a problem is not generally easy. One possible approach is to try to obtain $p(\theta)$ from the inversion of (7.14). This raises the topic of identifiability of mixtures of distributions and reference may be made for example to Teicher (1960, 1961, 1963). Difficulties arise over the uniqueness of the solution. For a given $p(y|\theta)$ and $p(y)$ there will not in general be a unique $p(\theta)$ which induces $p(y)$. This is true for example where the class of density functions $p(\cdot|\theta) : \theta \in \Theta\}$ is the class of normal distributions or of gamma distributions.

If one can satisfactorily obtain $p(\theta)$, however, the way is clear through the use of Bayes's theorem and

$$p(y|x) = \int_{\Theta} p(y|\theta)\, p(\theta|x) d\theta$$

to the determination of $p(y|x)$.

The general implications of starting with a prior $p(y)$ instead of a prior $p(\theta)$ can in some cases be illustrated neatly. Consider for example the case in which $p(y|\theta)$ is $\mathrm{Po}(\theta)$, and so $\mathrm{E}(y|\theta) = \mathrm{V}(y|\theta) = \theta$. We then have that

$$\mathrm{E}(y) = \mathrm{E}\{\mathrm{E}(y|\theta)\} = \mathrm{E}(\theta),$$

and $$\mathrm{V}(y) = \mathrm{E}\{\mathrm{V}(y|\theta)\} + \mathrm{V}\{\mathrm{E}(y|\theta)\} = \mathrm{E}(\theta) + \mathrm{V}(\theta).$$

From these relationships comparisons may be made concerning the corresponding choices of $p(y)$ or $p(\theta)$. The choice of a $p(y)$ which is concentrated about its mean corresponds to a choice of $p(\theta)$ similarly concentrated about its mean. At the other extreme the case where little is known about $p(y)$ with $V(y)$ large corresponds to the idea of prior ignorance on θ with $V(\theta)$ large.

7.6 Empirical Bayes prediction

In many situations we entertain the idea of the parameter θ being a random variable in the sense that in repetitions of the experiment different values of θ may be the true parameter value. In some situations, however, we may feel that we are not in a position to specify the distribution of θ – the randomness alone is not sufficient. Nor are we willing to assign a uniform or diffuse prior to θ to represent ignorance, which is in fact a very specialised assumption.

Although von Mises (1942) had discussed a similar topic the first real approach to such problems was by Robbins (1955), and it was he who coined the phrase 'empirical Bayes approach'. The assumption that θ is a random variable imposes a plausibility function $p(\theta)$ on Θ, but the randomness does not indicate the actual form of the distribution. This marginal distribution is therefore assumed unknown.

Robbins laid the foundations of the approach for obtaining the Bayes estimate of the parameter θ in the 'future' experiment and found in several instances that a direct estimate of $p(\theta)$ is not necessary. Instead attention is focussed on the marginal density function

$$p(y) = \int_{\Theta} p(y|\theta)p(\theta)\mathrm{d}\theta, \tag{7.15}$$

and it is sufficient simply to find an estimate of this from $x_1, x_2, \ldots, x_n$.

We can adapt the empirical Bayes approach to our problem of prediction. A sequence of experiments yields results $x_1, x_2, \ldots, x_n$. We do not know the values $\theta_1, \theta_2, \ldots, \theta_n$ of the parameters in the experiments but each is assumed to be an observation of a random variable θ with unknown probability density function $p(\theta)$. We require an informative predictor a for the future observation y in the next experiment, in which the unknown parameter value is θ_{n+1}. Using a quadratic utility function $-(a-y)^2$ we find that the expected utility is given by

$$-\int_Y (a-y)^2\, p(y)\mathrm{d}y. \tag{7.16}$$

This is maximised when a is the mean value of $p(y)$. If $p(\theta)$ is unknown we are unable to evaluate $p(y)$ and so cannot obtain the Bayes decision rule. In the empirical Bayes approach we use the results $x_1, x_2, \ldots, x_n$ to obtain an estimate $\hat{p}(y|\mathrm{x})$ of $p(y)$. For a discrete $p(y)$ we estimate $p(y)$ by $f_n(y)/n$, where $f_n(y)$ denotes the number of observations in the informative experiment having the value y. It follows immediately that the expected value of $\hat{p}(y|\mathrm{x})$ is simply the sample mean $\bar{x}$ of the informative experiment. For a continuous density $p(y)$ we would obtain our estimate $\hat{p}(y|\mathrm{x})$ in the form of a histogram, and approximate the expected value of $p(y)$ by the sample mean.

As in the Robbins estimation case mentioned above we see that it is unnecessary to obtain an approximation for $p(\theta)$. We focus our attention instead on the marginal density $p(y)$ and obtain an empirical estimate $\hat{p}(y|\mathrm{x})$. The difficulty presented by the mixture problem in the similar approach of §7.5 is thus avoided.

One severe assumption that is sometimes made is that the form of the prior distribution is known, but the particular member of the family is unknown. Consider for example the empirical Bayes approach to the following situation. Suppose that we have an independent set of observations $x_1, x_2, ..., x_n$ where $p(x_i|\theta) = \mathrm{No}(\theta, 1)$ and for the future experiment $p(y|\theta) = \mathrm{No}(\theta, 1)$. We assume that $p(\theta) = \mathrm{No}(b, c)$ where b, c are unknown. We have that

$$p(y) = \mathrm{No}\left(b, \left(1 + \frac{1}{c}\right)^{-1}\right).$$

We then obtain estimates of b, c from the results $x_1, x_2, ..., x_n$ of the informative experiment. For example we may take here

$$\hat{b} = \bar{x}, \hat{c} = \left(\frac{v}{n-1} - 1\right)^{-1},$$

where

$$\bar{x} = \frac{1}{n}\sum x_i, \quad v = S(\mathrm{x}, \mathrm{x}),$$

provided that $v > n - 1$. Hence our estimated predictive distribution is

$$\hat{p}(y|\mathrm{x}) = \mathrm{No}(\bar{x}, (n-1)v^{-1}).$$

Further references to the empirical Bayes approach may be found in Maritz (1970).

7.7 Distribution-free prediction

We conclude this chapter by taking an even more extreme starting point. Not only are we unable to assign a prior plausibility function – we cannot even identify the parameters or the form of distribution describing the informative and future experiments. The problem of whether or not to use a parametric form of density function to describe the future and informative experiment is one which requires very careful consideration and to which statisticians have probably paid too little attention. The art of approximation in the whole of applied mathematics is a subtle one, and the only justification for an approximation must be that it is of sufficient practical validity, or alternatively it is robust against departures from approximating assumptions under which it was derived. In statistical work the need to make an assumption about the functional form of the density may occasionally be obviated by the use of what have come to be termed non-parametric or distribution-free techniques. In prediction analysis these lead to the use of order statistics in frequentist inference prediction. The presentation of such results within our general framework is the main purpose of this section. Non-parametric methods are at present outside the reach of Bayesian techniques since there exists no satisfactory way of specifying measures on spaces of functions. There is, however, some Bayesian justification for the use, when there is a state of 'prior ignorance', of the empirical distribution as the predictive distribution $p(y|x)$, and for such cases it is of course then relatively easy to obtain the corresponding predictions.

Preliminary distribution theory. We first set out some definitions and properties of Dirichlet and beta distributions which will be required.

A Dirichlet density Di(g, h) is defined on the simplex

$$\left\{(u_1, \ldots, u_d): u_i \geqslant 0 \; (i = 1, \ldots, d), \sum_1^d u_i \leqslant 1\right\}$$

in d-dimensional Euclidean space by

$$\frac{1}{D(\mathrm{g}, h)} u_1^{g_1-1} u_2^{g_2-1} \ldots u_d^{g_d-1} (1-u_1-\ldots-u_d)^{h-1}. \qquad (7.17)$$

The special case where $d = 1$ is the more familiar beta type of distribution over the interval (0, 1).

Note that $\mathrm{E}(u_i) = g_i/(g_1 + \ldots + g_d + h) \quad (i = 1, \ldots, d)$. (7.18)

The first important property of Dirichlet distributions is that certain condensations preserve the Dirichlet form in the following way. Let $t_1, t_2, \ldots, t_j$ denote the sums of j non-overlapping subsets of $u_1, \ldots, u_d$; these sets need not necessarily exhaust the us. Then let $\mathrm{k} = (k_1, k_2, \ldots, k_j)$ denote the sums of the corresponding subsets of $g_1, \ldots, g_d$, and $l = g_1 + \ldots + g_d + h - k_1 - \ldots - k_j$. Then $(t_1, \ldots, t_j)$ has a Di(k, l) density. In particular $u_1 + \ldots + u_d$ has a $\mathrm{Di}(g_1 + \ldots + g_d, h)$ density.

Also of importance in distribution-free prediction analysis is a related *ordered* Dirichlet distribution Di*(g, h). This can be derived from $(u_1, \ldots, u_d)$ by the transformation

$$t_1 = u_1,$$
$$t_2 = u_1 + u_2,$$
$$\dots$$
$$t_d = u_1 + u_2 + \dots + u_d,$$

and gives as density, over the *ordered simplex*

$$\{(t_1, \dots, t_d): 0 \leqslant t_1 \leqslant t_2 \leqslant \dots \leqslant t_d \leqslant 1\}$$

in *d*-dimensional Euclidean space,

$$\frac{1}{D(\mathrm{g}, h)} t_1^{g_1 - 1} (t_2 - t_1)^{g_2 - 1} \dots (t_d - t_{d-1})^{g_d - 1} (1 - t_d)^{h-1}. \tag{7.19}$$

The important property for this distribution, and the one which corresponds to the property quoted for the Di distribution, is the following. If $(s_1, \dots, s_j)$ is any subset of $(t_1, \dots, t_d)$ and $\mathrm{k} = (k_1, \dots, k_j)$ gives the corresponding g_i values, then $(s_1, \dots, s_j)$ has a $\mathrm{Di}^*(\mathrm{k}, l)$ density, where

$$l = g_1 + \dots + g_d + h - k_1 - \dots - k_j.$$

Now we quote the basic distribution property of order statistics on which distribution-free prediction depends. Suppose that $x_1, \dots, x_n$ are the outcomes of n replicates of an experiment described by some *continuous* distribution function F, and that the ordered outcomes are $x_{(1)}, \dots, x_{(n)}$. Let $t_i = F(x_{(i)})$. Then the distribution of $(t_1, \dots, t_n)$ is $\mathrm{Di}^*(1', 1)$ and it is the non-dependence of this distribution on F which permits the derivation of distribution-free techniques. It then follows immediately that $(u_1, u_2, \dots, u_n)$ defined by

$$u_i = t_i - t_{i-1} \quad (i = 1, 2, \dots, n)$$

has a $\mathrm{Di}(1', 1)$ density.

Distribution-free tolerance intervals. Suppose that the informative experiment is n replicates of the future experiment and that associated with the future experiment is a distribution function F. We shall consider what are the consequences of using as either a mean coverage or a guaranteed coverage tolerance interval the interval $(x_{(r)}, x_{(s)})$, where $x_{(r)}$ and $x_{(s)}$ are the rth and sth order statistics of set $x = (x_1, \dots, x_n)$ of outcomes of the n-replicate experiment. The direction of investigation will be to determine the c values corresponding to the use of this interval (i) for the mean coverage approach and (ii) for the guaranteed coverage approach for a specified g value. Having found these c values we can then attempt to find r and s which provide satisfactory c values.

Mean coverage. In the mean coverage analysis we wish to obtain the expectation of

$$P_f\{(x_{(r)}, x_{(s)})\} = F(x_{(s)}) - F(x_{(r)}),$$

expectation being with respect to the n-replicate experiment (or its condensation in terms of $x_{(r)}, x_{(s)}$). Now, in fact, in the notation introduced,

$$\begin{aligned} F(x_{(s)}) - F(x_{(r)}) &= t_s - t_r \\ &= u_{r+1} + \ldots + u_s, \end{aligned}$$

and so has a $\mathrm{Di}(s-r, n+1-(s-r))$ distribution. Hence, from (7.18) its expectation is $(s-r)/(n+1)$. Thus to obtain a c mean coverage tolerance interval we have to attempt to choose r and s so that

$$c = \frac{s-r}{n+1}. \tag{7.20}$$

It is immediately clear, as we might have expected from the very limited class of predictors which we are considering, that by no means all values of c are possible, and it is therefore sensible to select (r, s) if possible in such a way that the chosen c value is exceeded. If a finite prediction interval is required we cannot obtain a mean cover value greater than $(n-1)/(n+1)$, and $r = 1$ and $s = n$ provide the maximum value of c. Similar considerations apply to infinite intervals where we take $r = 0$ and $x_{(0)} = -\infty$ or $s = n+1$ and $x_{(n+1)} = +\infty$. For finite intervals it is a common practice to take 'symmetric' intervals, i.e., with $s = n - r + 1$. Thus $(x_{(r)}, x_{(n-r+1)})$ supplies a tolerance interval of mean coverage $(n-2r+1)/(n+1)$.

A more general result may be stated. If we term the interval between consecutive order statistics (including conventionally $x_{(0)} = -\infty$ and $x_{(n+1)} = +\infty$) a *block*, then we may say that the region consisting of b such blocks provides a tolerance region of mean coverage $b/(n+1)$.

Guaranteed coverage. The c values corresponding to the use of $(x_{(r)}, x_{(s)})$ as a (c, g) guaranteed coverage tolerance interval is given by

$$P_e\{x: F(x_{(s)}) - F(x_{(r)}) \geqslant c\} = g. \tag{7.21}$$

Now since $F(x_{(s)}) - F(x_{(r)})$ has a $\mathrm{Di}(s-r, n+1-(s-r))$ density this is given by

$$\begin{aligned} &\int_c^{\infty} \frac{u^{s-r-1}(1-u)^{n+1-(s-r)-1}}{\mathrm{B}\{s-r, n+1-(s-r)\}}\, du \\ &= 1 - I_c\{s-r, n+1-(s-r)\}. \end{aligned}$$

For a given g and given c the problem is therefore to select r and s such that

$$1 - I_c\{s - r, n + 1 - (s - r)\} = g. \tag{7.22}$$

Again it is not necessarily possible to choose r and s so that c is exactly obtained and it is then necessary to choose so that c is just exceeded. This, of course, may prove to be impossible. It is in fact useful before conducting an informative experiment to know whether a desired c is attainable. From (7.22) it is clear that c will be attainable with a finite interval provided

$$1 - I_c(n - 1, 2) > g.$$

For given g and c it is easy to obtain the minimum n to meet the requirements from tables of incomplete beta function, or from special tables (Owen, 1962) constructed for the direct determination of n.

We now consider two examples illustrating the distribution-free approach for tolerance intervals. In the first example we investigate what, if anything, is lost by following the above approach when an assumption of normality in the underlying distribution is reasonable. In the second we consider a less standard situation.

Example 7.1

Crop Prediction. Recall example 5.4 on crop prediction. It is known that a normality assumption is reasonable. Consider however the distribution-free approach. The maximum value of c we can obtain with a two-sided interval is 0.92 from (7.20), and this is provided by the interval (4.1, 11.4). This is simply the range of the observations. The corresponding symmetric mean coverage interval obtained from (5.51) is (5.01, 10.93). Notice the widening in the interval when the normality assumption is not made.

We turn now to the guaranteed coverage tolerance interval. Again consider the interval (4.1, 11.4). This provides cover 0.847 with guarantee 0.90. The corresponding interval obtained by Howe's technique (see §6.5) is (6.0, 9.9).

Example 7.2

Component production. In the manufacture of a certain type of component it is possible for an airlock to get into the component. Such a component tends to suffer a reduction in its lifetime. Unfortunately it is not possible to detect the affected components. The lifetimes (in days) obtained for 185 components are summarised in the histogram in fig. 7.1, the ordered sample containing values (29.0, 30.7, 31.0, ..., 92.1, 92.6, 92.7, 93.2, 94.1). The distribution of lifetimes appears to be bimodal. A parametric approach would be to consider a mixture of two normal distributions, say, corresponding to the two types of component. This would involve problems in estimating the mixing parameter for use in the prediction of future lifetimes and lead to

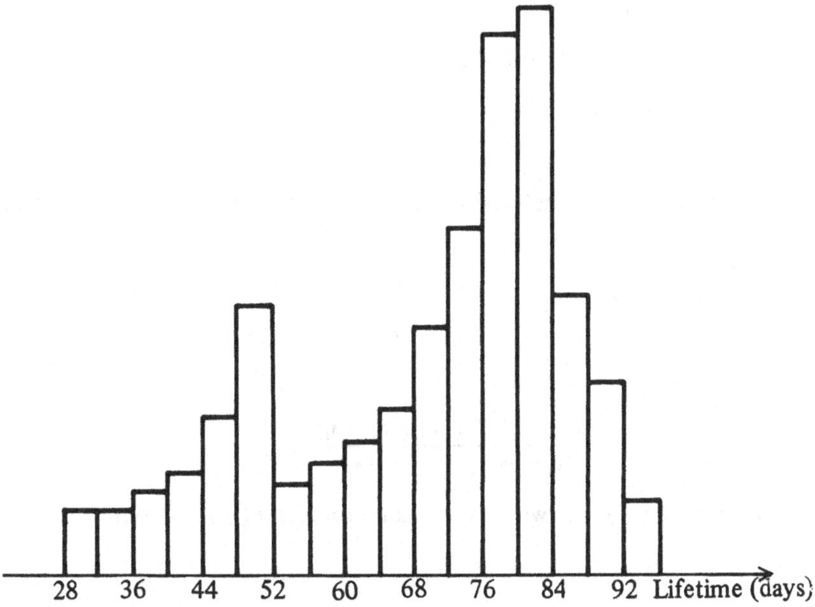

Fig. 7.1 Histogram of lifetimes of components in example 7.2.

non-standard situations which we have not covered. The distribution free approach here provides a simple alternative. For example the interval (31.0, 92.7) is a frequentist tolerance interval of mean coverage $c = 0.968$. Similarly this interval provides cover 0.986 with guarantee 0.95.

History

The frequentist decision theory approach and the similarity between linear utility and informative prediction intervals are discussed in Aitchison and Sculthorpe (1965) and Aitchison (1966).

The empirical Bayes approach was introduced by Robbins (1955) and heralded as a breakthrough by Neyman (1962). Maritz (1970) provides a good introduction to, and survey of, the topic and supplies a comprehensive bibliography.

The distribution-free approach to tolerance predictors is given by Wilks (1941) in the first important paper on tolerance regions. Guttman (1970, chapter 2) derives the relevant theory and provides several references.

Problems

7.1 For the case of the gamma distribution with

$$p(x|\theta) = \mathrm{Ga}(k, \theta),$$

$$p(y|\theta) = \mathrm{Ga}(K, \theta),$$

show that a frequentist linear utility interval of the form $\delta(x) = (0, qx)$ based on the utility function

$$U(\delta(x), y) = \begin{cases} -\xi(qx-y) & (y < qx), \\ -\eta(y-qx) & (y \geqslant qx), \end{cases}$$

is given by

$$q = \frac{1 - \mathrm{Be}\{k+1, K; \xi/(\xi+\eta)\}}{\mathrm{Be}\{k+1, K; \xi/(\xi+\eta)\}}.$$

Compare this interval with the tolerance predictor (5.40) of similar mean coverage c.

7.2 For the gamma case with

$$p(x|\theta) = \mathrm{Ga}(k, \theta), \; p(y|\theta) = \mathrm{Ga}(K, \theta),$$

find the frequentist linear utility interval of the form $\delta(x) = (q_1 x, q_2 x)$ and based on the utility function

$$U\{\delta(x), y\} = \begin{cases} \xi(y - q_1 x) & (y \leqslant q_1 x), \\ q_1 x - y & \{q_1 x < y \leqslant \tfrac{1}{2}(q_1 x + q_2 x)\}, \\ y - q_2 x & \{\tfrac{1}{2}(q_1 x + q_2 x) < y \leqslant q_2 x\}, \\ \lambda(q_2 x - y) & (y > q_2 x). \end{cases}$$

7.3 Suppose that

$$p(y|\theta) = \mathrm{Ex}(\theta), \; p(y) = \mathrm{InBe}(1, g, h).$$

What can be deduced about $p(\theta)$? If further

$$p(x|\theta) = \mathrm{Ga}(k, \theta)$$

what can be deduced about $p(y|x)$?

7.4 Derive an empirical Bayes predictive distribution for each of the following cases.

(i)
$$p(x_i|\theta) = \text{Po}(\theta) \qquad (i = 1, 2, ..., n),$$
$$p(y|\theta) = \text{Po}(\theta),$$
$$p(\theta) = \text{Ga}(g,h) \quad \text{with } g, h \text{ unknown.}$$

(ii)
$$p(x_i|\theta) = \text{Ex}(\theta) \qquad (i = 1, 2, ..., n),$$
$$p(y|\theta) = \text{Ex}(\theta),$$
$$p(\theta) = \text{Ga}(g,h) \quad \text{with } g, h \text{ unknown.}$$

7.5 In an effort to cut costs a builder is considering whether he can install smaller domestic storage tanks than the ones he uses at present in his houses. He wishes to decide the minimum size of tank that he can reasonably install in his standard 3-bedroomed houses. The water supply to the storage tank is through a valve which opens whenever water is drawn off and closes when the level has been restored to the 'Full' mark. The supply rate is constant when in use and is such that for the domestic demand pattern he is considering the tanks at present in use never became empty and each day began at the 'Full' mark. For a random selection of his standard houses he monitors the water usage by measuring the maximum drop in level from the 'Full' mark on 365 days. The results set out as an ordered sample consist of values 6.1, 6.5, 6.5, ..., 47.5, 50.1, 54.3, 55.9, 57.1 litres. Can you assist the builder to reduce his storage tank size?

8
Sampling inspection

8.1 Introduction

The myriad of possible statistical sampling inspection procedures forces us to consider in detail only a very small selection in a book of this size. We would need a separate book to do justice to the huge variety of plans. In this chapter therefore we show how some standard plans come within the framework of decisive prediction, and how the framework can readily cope with less standard problems. The application of prediction theory to this area will provide some additional justification and motivation for some of these plans. We hope that those selected will be sufficient to indicate the direction of analysis to any reader with a specific problem.

We consider both fixed size sample and sequential sampling schemes. Wetherill (1966) and Wetherill and Campling (1966) also provide a decision theory approach to sampling inspection but do not consider predictive distributions.

8.2 Fixed-size single-sample destructive testing

We consider first a fixed-size single-sample plan for deciding whether to accept or reject a batch. For a process which produces an item at each of a number of independent operations we may imagine as our basic future experiment the determination of the quality y of a single item. This quality y may be a simple counting variable taking the value 1 for an effective and 0 for a defective item, or may be more sophisticated, for example the lifetime of a component or the degree of purity of a chemical preparation. We suppose that the probabilistic mechanism which describes the production of the variable y is a density function $p(y|\theta)$ on Y where, as in previous work, θ is an indexing parameter with density function $p(\theta)$. We suppose further that altogether N items have been produced by independent operations of the system and that the informative experiment consists of the *destructive* testing of n components, whose qualities $x_1, \ldots, x_n$ are determined by the testing. The informative experiment e is then described by the density function

$$p(\mathrm{x}|\theta) = p(x_1|\theta) \ldots p(x_n|\theta).$$

The utility structure is such that the realised utility depends on the as yet unknown qualities $y_1, \ldots, y_{N-n}$ of the untested items. It also depends on which of the available possible actions is adopted. For simplicity we suppose that the action space A consists of just two possible actions:

$$\begin{aligned} a_1 &: \text{ market the untested items,} \\ a_2 &: \text{ scrap the untested items.} \end{aligned} \tag{8.1}$$

We can thus express the utility in the form $U(a, y_1, \ldots, y_{N-n})$ or $U(a, \mathbf{y})$. Note that we have not made allowance in this basic utility for the cost of testing the items. This is not necessary at the terminal stage of determining the appropriate action after the testing is complete. It is, however, very relevant to the preposterior analysis for deciding how many items should be tested, and will appear in §8.4.

The inferential aspect of the problem produces the predictive density function

$$p(\mathbf{y}|\mathbf{x}) = \int_\Theta p(y_1|\theta) \ldots p(y_{N-n}|\theta)\, p(\theta|\mathbf{x})\mathrm{d}\theta. \tag{8.2}$$

Note that the distribution of $y_1, \ldots, y_{N-n}$ for given $\mathbf{x}$ is not necessarily a product distribution. In general the function $p(\mathbf{y}|\mathbf{x})$ will not be the product of component functions $p(y_1|\mathbf{x}), \ldots, p(y_{N-n}|\mathbf{x})$.

The optimum action is then easily decided. According to standard statistical decision theory we should take that action which maximises

$$U(a) = \int_\mathbf{Y} U(a, \mathbf{y})\, p(\mathbf{y}|\mathbf{x})\, \mathrm{d}\mathbf{y}, \tag{8.3}$$

where the bold $\mathbf{dy}$ and $\mathbf{Y}$ indicate that integration is over the $(N-n)$-dimensional space, the product space of $N-n$ identical spaces $Y_1, \ldots, Y_{N-n}$, each of type Y.

A great simplification occurs if the utilities are additive over items; in other words, if we can envisage a component utility attaching to each item, this utility depending only on the quality of the item, and if the utility of the batch of $N-n$ items is the sum of these component utilities. Mathematically, if $U(a, y)$ denotes the utility of a component of quality y then

$$U(a, \mathbf{y}) = U(a, y_1) + \ldots + U(a, y_{N-n}). \tag{8.4}$$

Such a utility function would of course not apply if the utility of the batch depended on some overall property of the batch, for example, if it were of value only if it contained at most d defective items.

The simplification with utility function (8.4) arises because we can rewrite (8.3) as

$$U(a) = \int_{\mathbf{Y}} \int_{\Theta} \sum_{i=1}^{N-n} U(a, y_i)\, p(y_1|\theta) \ldots p(y_{N-n}|\theta)\, p(\theta|\mathbf{x})\, \mathrm{d}\mathbf{y}\mathrm{d}\theta. \tag{8.5}$$

The contribution from the first term of the sum can (provided the change of order of operations is allowable) be written as

$$\begin{aligned} &\int_{Y_1} U(a, y_1) \mathrm{d}y_1 \int_{\Theta} p(y_1|\theta)\, p(\theta|\mathbf{x})\mathrm{d}\theta \\ &\times \int_{Y_2 \times \ldots \times Y_{N-n}} p(y_2|\theta) \ldots p(y_{N-n}|\theta) \mathrm{d}y_2 \ldots \mathrm{d}y_{N-n} \\ &= \int_{Y_1} U(a, y_1)\, p(y_1|\mathbf{x}) \mathrm{d}y_1. \end{aligned}$$

Hence

$$U(a) = (N-n) \int_{Y} U(a, y)\, p(y|\mathbf{x}) \mathrm{d}y. \tag{8.6}$$

Example 8.1

Batch acceptance. A manufacturer must decide whether to market or scrap a batch of N items which he has produced. By destructive testing he determines the nature $x_1, x_2, \ldots, x_n$ of a random sample of n items, each being classified as either effective ($x_i = 1$) or defective ($x_i = 0$). If $x = \sum_{i=1}^{n} x_i$ records the number of effective items a suitable description is a binomial trials model with $p(x|\theta) = \mathrm{Bi}(n, \theta)$. The acceptability of the remaining $N-n$ items in the batch for a customer depends on the characteristics $y_1, y_2, \ldots, y_{N-n}$ of the items, where

$$\left.\begin{aligned} p(y_i &= 1|\theta) = \theta \\ p(y_i &= 0|\theta) = 1-\theta \end{aligned}\right\} \quad (i = 1, 2, \ldots, N-n).$$

If we take $p(\theta) = \mathrm{Be}(g, h)$ it follows from table 2.3 and (8.2) that the predictive distribution for $\mathbf{y} = (y_1, y_2, \ldots, y_{N-n})$ is given by

$$p(\mathbf{y}|x) = \frac{\mathrm{B}(G + \Sigma y_i, H + N - n - \Sigma y_i)}{\mathrm{B}(G, H)}$$

$$(\mathbf{y}: y_i = 0, 1;\ i = 1, 2, \ldots, N-n), \tag{8.7}$$

where $G = g + x$, $H = h + n - x$.

Suppose that the customer considers the batch effective only if there are at least q effective items remaining, and that the manufacturer provides a 'money-back' guarantee if the batch does not meet this specification. Then for the two actions (8.1) the manufacturer assesses his gains as follows:

$$U(a_1, y) = \begin{cases} K_1 & \text{if } \sum_{1}^{N-n} y_i \geq q, \\ -K_2 & \text{if } \sum_{1}^{N-n} y_i < q, \end{cases}$$
$$U(a_2, y) = -L, \tag{8.8}$$

where K_1, K_2, L are non-negative constants. We then have

$$U(a_1) = (K_1 + K_2) \sum_{i=0}^{N-n-q} \binom{N-n}{i} \frac{B(G+N-n-i, H+i)}{B(G,H)} - K_2,$$

and

$$U(a_2) = -L. \tag{8.9}$$

We select the action giving $\max\{U(a_1), U(a_2)\}$. Notice that we need investigate only the case $L < K_2$; for otherwise the optimal action would be to market every batch irrespective of the x-value, thus neglecting any information obtained from sampling.

Consider now a numerical solution. Suppose that $N = 100$, $n = 10$ and $q = 80$. Also suppose that the manufacture and testing of the batch costs £40 and the selling price of the batch is £60. Thus $K_1 = 20$ and $K_2 = 40$. If the batch is rejected the manufacturer can sell it for scrap for £1; hence $L = 39$. The manufacturer has been operating the process for some time and assesses his knowledge of θ by $p(\theta) = \text{Be}(9, 1)$. This gives a prior mean success rate of 0.90 and, for example, $P(\theta > 0.7) = 0.96$. This indicates that the manufacturer has fairly strong prior ideas about θ.

The values of $U(a_1)$ for the possible x-values are shown in table 8.1. Our optimal action is to market the batch if we get 5 or more effective items in the sample; otherwise scrap the batch. Notice how critical the scrapping loss is in this situation.

The manufacturer here has a good deal of prior information on θ and this affects his decision rule to a considerable degree. He is able to allow up to 5 defective items in a sample of 10 before he scraps the batch because he is so confident that θ will be large; and the probability of obtaining as low an x-value as 5 consequently will be very small. If the process is new and the manufacturer thinks any value of θ equally plausible, so that $p(\theta) = \text{Be}(1, 1)$, the corresponding values of $U(a_1)$ are also shown in table 8.1, and the optimal action is to market the batch only if 7 or more effective items are found.

The expected utilities shown in table 8.1 may seem somewhat strange and indicate that the manufacturer has little time left before going out of business!

Table 8.1 *Values of* $U(a_1)$, *the expected utility (8.9) of batch acceptance with* $K_1 = 20$, $K_2 = 40$, *and of* $p(x)$, *the prior predictive density (8.10), corresponding to the two cases* Be*(9, 1) and* Be*(1, 1) for* $p(\theta)$

x	Be(9, 1) $U(a_1)$	Be(9, 1) $p(x)$	Be(1, 1) $U(a_1)$	Be(1, 1) $p(x)$
0	−40.00	0.000	−40.00	1/11
1	−39.99	0.000	−40.00	1/11
2	−39.95	0.000	−40.00	1/11
3	−39.82	0.002	−40.00	1/11
4	−39.36	0.005	−39.98	1/11
5	−38.01	0.014	−39.89	1/11
6	−34.67	0.033	−39.40	1/11
7	−27.68	0.070	−37.37	1/11
8	−15.83	0.139	−31.06	1/11
9	− 0.38	0.263	−16.82	1/11
10	13.49	0.474	4.29	1/11

Before he performs his experiment his ideas about the value of x he is likely to obtain are summarised by the prior predictive distribution

$$p(x) = \int_{\Theta} p(x|\theta)\, p(\theta) d\theta. \tag{8.10}$$

Here this is given by BeBi$(10, g, h)$. If the manufacturer follows his optimal choice he will expect to gain

$$\sum_{x=0}^{10} \max\{U(a_1), U(a_2)\} p(x). \tag{8.11}$$

For the two prior functions considered, namely Be(9, 1) and Be(1, 1) the prior predictive densities $p(x)$ are shown in table 8.1, and the expected utilities (8.11) are 0.20 and −32.18 respectively. The foolhardiness of the manufacturer's venture is amply demonstrated in the latter case.

The above serves as an introduction to preposterior analysis. We return to this concept in §8.4.

Suppose that instead of selling the remainder of the batch as a whole, the items are sold individually with a double-your-money back guarantee for an item which is defective. The utility to the manufacturer may then be different and a sensible assessment would be of the form (8.4) with components

$$\left.\begin{aligned} U(a_1, y_i) &= \begin{cases} k_1 & \text{if } y_i = 1, \\ -k_2 & \text{if } y_i = 0, \end{cases} \\ U(a_2, y_i) &= -l, \end{aligned}\right\} \tag{8.12}$$

($i = 1, 2, ..., N-n$), where k_1, k_2, l are non-negative constants. Thus if he sells an effective item he will make a profit of k_1 per item; if he markets a defective item he has to return twice the selling price with a corresponding loss of k_2 per item. Again there is a scrapping loss, assessed as l per scrapped item. Thus each individual item makes its own independent contribution to the overall utility of the action. Now

$$p(y_i|x) = \begin{cases} \dfrac{G}{G+H} & \text{if} \quad y_i = 1, \\ \dfrac{H}{G+H} & \text{if} \quad y_i = 0, \end{cases} \tag{8.13}$$

and so

$$\left.\begin{aligned} U(a_1) &= \frac{(N-n)(k_1G-k_2H)}{G+H} = (N-n)\left\{\frac{(k_1+k_2)G}{G+H} - k_2\right\}, \\ U(a_2) &= -(N-n)l. \end{aligned}\right] \tag{8.14}$$

Here it is sensible to assume $l < k_2$.

The determination of the optimal action follows directly. Consider again a numerical solution with $N = 100$, $n = 10$. Suppose that each item costs £0.40 to produce and is sold at £0.60; hence $k_1 = 0.2, k_2 = 1$. A rejected item may be sold for scrap for £0.05, giving $l = 0.35$. (These figures are not necessarily compatible with the values for K_1, K_2, L assumed earlier.) Again we take $p(\theta) = \text{Be}(9, 1)$. Table 8.2 provides the values of $U(a_1)$ for the possible outcomes x of the informative experiment. The optimal terminal decision rule is therefore the following.

If $x = 0, 1$, scrap the batch;
otherwise, market the batch.

Table 8.2 *Values of $U(a_1)$, the expected utility (8.14) for batch acceptance with $k_1 = 0.2$, $k_2 = 1.0$, corresponding to the two cases* Be(9,1) *and* Be(1, 1) *for $p(\theta)$*

x	$U(a_1)$ corresponding to $p(\theta) =$ Be(9, 1)	Be(1, 1)
0	−41.4	−81
1	−36.0	−72
2	−30.6	−63
3	−25.2	−54
4	−19.8	−45
5	−14.4	−36
6	− 9.0	−27
7	− 3.6	−18
8	1.8	− 9
9	7.2	0
10	12.6	9

This may seem a somewhat unlikely optimal rule. As before, the prior distribution is playing a dominant role – we are so sure θ will be large (and consequently the batch will be acceptable) that only the most extreme values of x will cause us to scrap the batch. Again as comparison we list the values of $U(a_1)$ in table 8.2 for the uniform prior $p(\theta) = \text{Be}(1, 1)$. The terminal decision rule for this prior would be the following.

If $x = 0, 1, ..., 5$, scrap the batch;
otherwise, market the batch.

8.3 Role of mean coverage and guaranteed coverage tolerance predictors in sampling plans

The decision-theoretic approach of §8.2 to sampling inspection throws some light on commonly suggested sampling plans in quality control. Such sampling plans usually make the demand that for a batch to be marketable some statistical tolerance limit constructed from a random sample from the batch should meet some quality requirement such as the exceeding of some specified quality.

Mean coverage predictors. Suppose that there is a critical quality level q for each item. For any marketed item with quality level q or more there is a profit of k_1, whereas for any marketed item with quality level below q there is a loss of k_2. The loss involved in scrapping a batch is l per item. Thus we can specify the utility structure (8.4) with components

$$\left.\begin{aligned} U(a_1, y_i) &= \begin{cases} k_1 \; (y_i \geqslant q), \\ -k_2 \; (y_i < q), \end{cases} \\ U(a_2, y_i) &= \; -l \qquad\qquad (i = 1, 2, ..., N-n). \end{aligned}\right\} \tag{8.15}$$

This is simply a generalisation of utility function (8.12) It follows that

$$\left.\begin{aligned} U(a_1) &= (N-n)\left\{(k_1+k_2)\int_q^\infty p(y\,|\mathrm{x})\mathrm{d}y \; - k_2\right\}, \\ U(a_2) &= -(N-n)\,l. \end{aligned}\right\} \tag{8.16}$$

We shall take a decision to market if and only if $U(a_1) > U(a_2)$, that is, if and only if

$$\int_q^\infty p(y|\mathrm{x})\mathrm{d}y > \frac{k_2 - l}{k_1 + k_2}. \tag{8.17}$$

Notice we again need to assume $l < k_2$ for a realistic situation. The inequality (8.17) can be expressed more familiarly by noting that it is equivalent to the lower Bayesian informative predictor of cover $(k_2 - l)/(k_1 + k_2)$ computed from the sample exceeding the critical quality q; in other words, if we used an informative prediction interval (q, ∞) for the quality of a future item the Bayesian cover provided by it would be at least $(k_2 - l)/(k_1 + k_2)$. The importance of this result is that it gives a tangible meaning of the cover associated with the sampling plan in terms of more directly assessable profits and losses; cf. §7.2.

Guaranteed coverage predictors. Suppose that the batch is large and that it is effective if and only if the proportion of items in the batch of quality q or more is at least c. Under such circumstances it is then necessary to employ the alternative utility formulation of §3.6. If the profit from an effective batch is K_1, the loss from a defective batch K_2, and the loss from a scrapped batch L then we can set

$$V(a_1, \theta) = \begin{cases} K_1 & \text{if } P_f\{(q, \infty)|\theta\} \geqslant c, \\ -K_2 & \text{otherwise.} \end{cases} \tag{8.18}$$

This is a generalisation of the situation covered by utility function (8.8). Then

$$\begin{aligned} U(a_1) &= \int_\Theta V(a_1, \theta)\, p(\theta|\mathrm{x}) d\theta \\ &= (K_1 + K_2) \int_{[\theta: P_f\{(q, \infty)|\theta\} \geqslant c]} p(\theta|\mathrm{x}) d\theta - K_2, \end{aligned} \tag{8.19}$$

and

$$\begin{aligned} U(a_2) &= \int_\Theta V(a_2, \theta)\, p(\theta|\mathrm{x}) d\theta \\ &= -L. \end{aligned}$$

Hence we should market if and only if

$$P[\theta: P_f\{(q, \infty)|\theta\} \geqslant c|\mathrm{x}] > \frac{K_2 - L}{K_1 + K_2}. \tag{8.20}$$

We can express this marketing rule very simply in terms of what may be called a Bayesian guaranteed coverage tolerance limit. In chapters 4 and 6 we did not define such an interval. However the extension to such a concept is obviously straightforward. On the basis of the sample information x construct a lower (c, g) Bayesian guaranteed coverage limit with $g = (K_2 - L)/(K_1 + K_2)$.

If the tolerance limit falls below the critical quality level q the batch should be marketed; otherwise it should be scrapped.

8.4 Optimum choice of fixed sample size

In our study of sampling plans so far we have considered only the terminal stage of the analysis, that is the problem of determining the appropriate action given the sample design and after the testing is complete. The choice of experiment or sampling design is obviously important and we now investigate this preposterior analysis.

We will again confine our attention to the fixed sample size destructive testing design of §8.2. The problem is to select the size n of the sample which we should take in the informative experiment. An extension of the notation is required. Suppose we rewrite $U(a, \mathrm{y})$ as $U(n, \mathrm{x}, a, \mathrm{y})$, that is the utility of drawing a sample of size n, observing outcomes $\mathrm{x} = (x_1, x_2, ..., x_n)$ and choosing action a when $\mathrm{y} = (y_1, y_2, ..., y_{N-n})$ is the outcome of the future experiment. The terminal analysis leads us to select the action which maximises

$$U(n, \mathrm{x}, a) = \int_{\mathbf{Y}} U(n, \mathrm{x}, a, \mathrm{y})\, p(\mathrm{y}|\mathrm{x})\mathrm{d}\mathrm{y}. \tag{8.21}$$

(Strictly we should indicate the dependence of $p(\mathrm{y}|\mathrm{x})$ on n by writing $p(\mathrm{y}|\mathrm{x}, n)$. However we shall retain our usual notation for the predictive density function.) Thus the expected utility of sampling n items and observing x is given by

$$U(n, \mathrm{x}) = \max_{a \in A} U(n, \mathrm{x}, a). \tag{8.22}$$

In a fixed sample size plan we must select n before experimentation. Although we do not know at that stage which x will obtain, we do have a distribution over the possible values given by

$$p(\mathrm{x}) = \int_{\Theta} p(x_1|\theta) \dots p(x_n|\theta)\, p(\theta)\mathrm{d}\theta. \tag{8.23}$$

We may therefore evaluate the expected utility of performing an informative experiment with sample size n, namely

$$U(n) = \int_{\mathbf{X}} U(n, \mathrm{x})\, p(\mathrm{x})\mathrm{d}\mathrm{x}. \tag{8.24}$$

This preposterior analysis supplies the optimal size of sample—the value of n which maximises $U(n)$.

Although no theoretical problems present themselves in the preposterior analysis, the actual determinations of the optimal sample size in a practical situation may be tedious. As an illustration we investigate the preposterior analysis for example 8.1.

Example 8.1 (continued)

Consider the case of selecting a sample of size n from the batch of N items, and observing $\mathbf{x} = (x_1, x_2, ..., x_n)$. Suppose that we consider the case where the remaining items are sold individually so that we may take as our utility function

$$U(n, x, a_j, \mathbf{y}) = \sum_{i=1}^{N-n} U(n, x, a_j, y_i) - \gamma n, \tag{8.25}$$

where $U(n, x, a_j, y_i)$ is given in (8.12) ($i = 1, 2, ..., N-n$) and where $x = \Sigma x_i$. The utility function now includes a factor γn, with $\gamma > 0$, which gives a measure of the cost of sampling n items. Thus γ may be termed the cost per item sampled. Inevitably this involves some compromise between the gains and losses accrued through the information obtained by destroying a defective item on the one hand and a good item on the other. From (8.21) we have the generalisation of (8.14):

$$\left.\begin{aligned} U(n, x, a_1) &= (N-n)\left\{\frac{(k_1+k_2)(g+x)}{g+h+n} - k_2\right\} - \gamma n, \\ U(n, x, a_2) &= -l(N-n) - \gamma n, \end{aligned}\right\} \tag{8.26}$$

so that the terminal analysis yields

$$U(n, x) = \begin{cases} U(n, x, a_1) & \text{if } n \geqslant x \geqslant Q, \\ U(n, x, a_2) & \text{otherwise,} \end{cases} \tag{8.27}$$

where

$$Q = \frac{(g+h+n)(k_2-l)}{k_1+k_2} - g. \tag{8.28}$$

Notice that if $Q \leqslant 0$ we will always take action a_1, and if $Q > n$ the action a_2.

The prior predictive distribution $p(x)$ given by (8.23) is BeBi (n, g, h). Substitution in (8.24) yields

$$U(n) = \begin{cases} \dfrac{(N-n)(k_1 g - k_2 h)}{g+h} - \gamma n & \text{if } Q \leqslant 0, \\ U^*(n) - \gamma n & \text{if } 0 < Q \leqslant n, \\ -l(N-n) - \gamma n & \text{if } Q > n, \end{cases} \tag{8.29}$$

where

$$U^*(n) = -(N-n)\left\{\frac{(k_1+k_2)g}{g+h+n} - k_2 + l\right\}\sum_{x=0}^{[Q]} p(x) + \frac{(N-n)(k_1+k_2)}{g+h+n}\sum_{x=0}^{[Q]} xp(x) + \frac{(N-n)(k_1 g - k_2 h)}{g+h}. \tag{8.30}$$

Table 8.3 *Values of U(n), the expected utility (8.29) of sample size n corresponding to the two cases* Be*(9, 1) and* Be*(1, 1) for p(θ)*

n	$U(n)$ corresponding to $p(\theta) =$ Be(9, 1)	Be(1, 1)
0	8.00	−35.00
1	$7.92-\gamma$	$-27.23-\gamma$
2	$7.84-2\gamma$	$-26.13-2\gamma$
3	$7.76-3\gamma$	$-24.74-3\gamma$
4	$7.68-4\gamma$	$-24.00-4\gamma$
5	$7.60-5\gamma$	$-23.41-5\gamma$
6	$7.52-6\gamma$	$-22.83-6\gamma$
7	$7.44-7\gamma$	$-22.48-7\gamma$
8	$7.36-8\gamma$	$-21.98-8\gamma$
9	$7.28-9\gamma$	$-21.72-9\gamma$
10	$7.20-10\gamma$	$-21.27-10\gamma$
11	$7.12-11\gamma$	$-21.02-11\gamma$
12	$7.04-12\gamma$	$-20.65-12\gamma$
13	$6.96-13\gamma$	$-20.38-13\gamma$
14	$6.88-14\gamma$	$-20.07-14\gamma$

If we take the same numerical example as in §8.2, with $N = 100, k_1 = 0.2, k_2 = 1, l = 0.35, g = 9, h = 1$, we find that, provided $\gamma > 0$, it is not worth while to sample the batch. The optimal value of n is 0. We have that for $n = 0, 1, 2, ..., 6$, $Q < 0$ anyway, and so in such cases action a_1 is optimal. The values of $U(n)$ are shown in table 8.3 for $n = 0, 1, ..., 14$, so that the overall expected gain to the manufacturer is £8.00.

For comparison we again look at the case where $p(\theta) = \text{Be}(1,1)$. Table 8.3 gives the corresponding values of $U(n)$. Suppose the manufacturer assesses $\gamma = 0.40$. We have that the optimal sample size is $n = 8$ and the batch is marketed only if 5 or more effective items are observed. Notice however that the manufacturer will always expect to lose, and so an investigation of his costs, his guarantee, his prior information and his plant is in order!

8.5 Sequential predictive sampling inspection

One possible alternative to the fixed-size, single-sample destructive testing plan for batch acceptance or rejection is a sequential scheme. Here we test one item at a time, and after each trial we assess the situation and select one of three possible actions, namely,

a_1 : market the untested items,
a_2 : scrap the untested items,
a_3 : test another item.

The utility structure will depend on the number $(N-n)$ of items left and we explicitly show this by writing the utility in the form $U(a, \mathbf{y}_{N-n})$, where $\mathbf{y}_{N-n} = (y_1, y_2, ..., y_{N-n})$. (Note that we have reverted to our shortened notation of §8.2, 8.3.) As in §8.4 there is a cost γ attached to testing an item.

Let $F_n(\mathbf{x}_n)$ denote the maximum expected gain from pursuing an optimal policy after n items have been tested with results $\mathbf{x}_n = (x_1, x_2, ..., x_n)$. Then the principle of optimality leads to the following relationships.

$$\left.\begin{aligned} F_n(\mathbf{x}_n) &= \max\left\{ \int_{Y_1 \times \ldots \times Y_{N-n}} \cdots \int U(a_1, \mathbf{y}_{N-n})\, p(\mathbf{y}_{N-n} | \mathbf{x}_n)\, d\mathbf{y}_{N-n}, \right. \\ &\qquad \int_{Y_1 \times \ldots \times Y_{N-n}} \cdots \int U(a_2, \mathbf{y}_{N-n})\, p(\mathbf{y}_{N-n} | \mathbf{x}_n)\, d\mathbf{y}_{N-n}, \\ &\qquad \left. \int_Y F_{n+1}\{(\mathbf{x}_n, y)\}\, p(y | \mathbf{x}_n)\, dy - \gamma \right\} \\ &\qquad (n = 0, 1, 2, ..., N-1), \\ F_N(\mathbf{x}_N) &= 0. \end{aligned}\right\} \tag{8.31}$$

Again considerable simplifications occur if the utilities are additive over items as in (8.4) of §8.2. For then

$$F_n(\mathbf{x}_n) = \max\left\{ (N-n) \int_Y U(a_1, y)\, p(y | \mathbf{x}_n)\, dy, \right.$$
$$\left. (N-n) \int_Y U(a_2, y)\, p(y | \mathbf{x}_n)\, dy,\ \int_Y F_{n+1}\{(\mathbf{x}_n, y)\}\, p(y | \mathbf{x}_n)\, dy - \gamma \right\}. \tag{8.32}$$

Example 8.1 (continued)

Suppose that we consider testing the batch of size N with a sequential sampling scheme and that utility specification (8.12) is appropriate. Then, writing

$x = \sum_1^n x_i$, we have from (8.32)

$$\left.\begin{aligned} F_n(x) &= \max\left[(N-n)\left\{\frac{(k_1+k_2)G}{G+H}-k_2\right\},\right. \\ &\qquad \left. -(N-n)l, \frac{H}{G+H}F_{n+1}(x)+\frac{G}{G+H}F_{n+1}(x+1)-\gamma\right], \\ &\qquad (n = 0, 1, ..., N-1; x = 0, 1, ..., n), \\ F_N(x) &= 0; \end{aligned}\right\} \tag{8.33}$$

with $G = g+x, H = h+n-x$. The optimal strategy must then be obtained by a standard dynamic programming technique.

To provide a numerical solution we take the case where $N = 20$, the other constants being as in §§8.2, 8.4, namely $k_1 = 0.2$, $k_2 = 1.0$, $l = 0.35$, $\gamma = 0.40$. Consider first the case where $p(\theta) = \text{Be}(1,1)$. From (8.33) we have

$$\left.\begin{aligned} F_n(x) &= \max\left[(20-n)\left\{\frac{1.2(1+x)}{n+2}-1\right\}, -0.35(20-n),\right. \\ &\qquad \left. \frac{n+1-x}{n+2}F_{n+1}(x)+\frac{x+1}{n+2}F_{n+1}(x+1)-0.40\right] \\ &\qquad (n = 0, 1, ..., 19;\ x = 0, 1, ..., n), \\ F_{20}(x) &= 0. \end{aligned}\right\} \tag{8.34}$$

Fig. 8.1 shows the optimal actions for each possible (n, x) position. The procedure is to start at the origin and continue sampling until a boundary is reached. The path followed moves at each step either horizontally along one square for a defective item or diagonally upwards across one square for an effective item.

For the case where $p(\theta) = \text{Be}(9,1)$ it turns out that the optimal action is to accept the batch without sampling; compare the similar situation in §8.4 with $N = 100$. For the more uncertain prior situation just considered it is worth while obtaining some information by testing before coming to a decision.

History

There is a vast literature on sampling inspection models, a not negligible proportion of which is concerned with Bayesian models; for example Guthrie and Johns (1959), Lindley and Barnett (1965) and Wetherill and Campling (1966). In the decision theoretic models the tendency is to work with utility

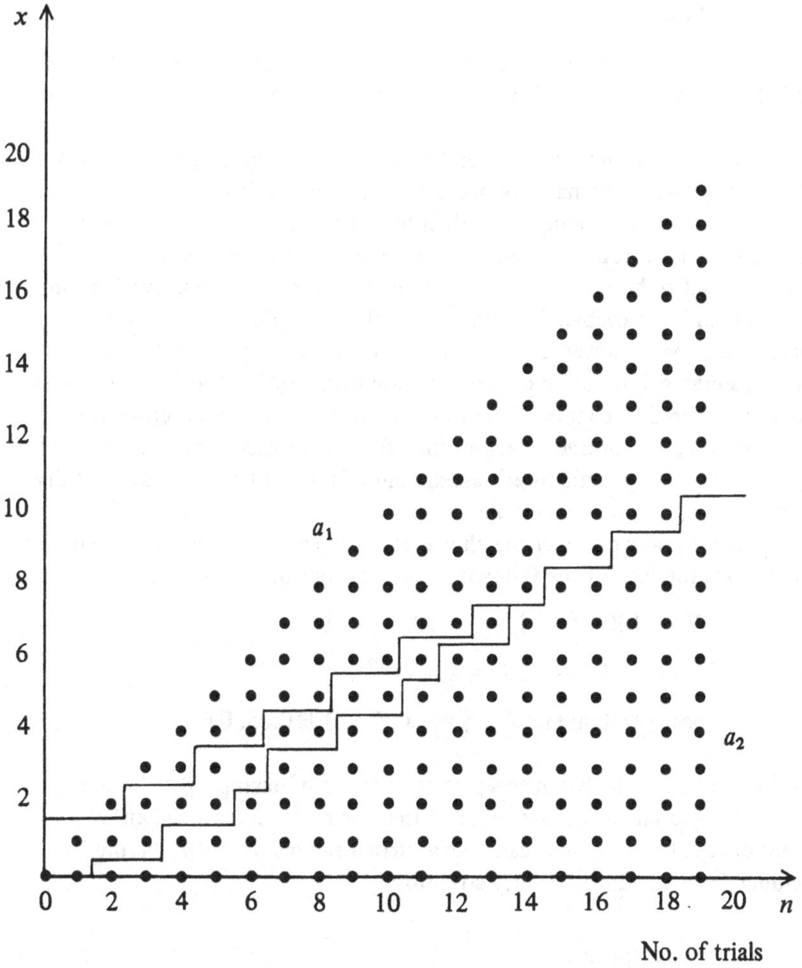

Fig. 8.1 Sequential plan for batch acceptance problem.

specifications of the form $V(a, \theta)$, see §3.6, and not to consider explicitly the concept of predictive distributions. Draper and Guttman (1968*a*, 1968*b*) use predictive distributions in their Bayesian model, albeit only for vague prior distributions.

Wetherill (1966) gives a good review of sequential decision models.

Problems

8.1 Follow through the terminal and preposterior analysis of Example 8.1 if $p(x|\theta)$ is $\text{No}(\mu, \tau)$ instead of binomial and if utility specification (8.15) holds.

8.2 A manufacturer must decide whether to market or destroy a batch of N machine tools which he has produced. The suitability of the machine tools in a batch for a customer depends on their lifetimes. The manufacturer gets some information from the destructive testing of n machine tools whose lifetimes turn out to be $x_1, x_2, \ldots, x_n$, these being independent observations on an $\text{Ex}(\theta)$ random variable. The lifetimes of the remaining machine tools remain unknown however. It is recognised that θ may vary from batch to batch, a suitable description of this variation being $p(\theta) = \text{Ga}(g, h)$. A machine tool is considered of beneficial use to the customer only if its lifetime is greater than q; the contract specifies therefore the replacement of non-useful components. The manufacturer's assessment of this contract yields the utility structure (8.15). Should he market the batch?

Suppose we now consider the choice of sample size n. Determine the optimal sample sizes for each of the following specifications of the constants:

$$N = 100, \; g = 1, \; h = 10, \; q = 1,$$

$$k_1 = 1, \; k_2 = 4, \; l = 0,$$

cost per item sampled $\gamma = 1.0, 0.5, 0.1, 0.05, 0.01$.

8.3 Consider again the batch acceptance situation of example 8.1. Suppose that the sample taken is of size n_1, and that the manufacturer can either accept or reject the batch or can take a further sample of size n_2 before deciding. With an additive utility structure

$$U(\text{accept}, y_i) = \begin{cases} k_1 & \text{if } y_i = 1, \\ -k_2 & \text{if } y_i = 0, \end{cases}$$

$$U(\text{reject}, y_i) = -l,$$

for the remaining items together with a cost γ per sampled item, which action would you take in this double-sampling scheme if the first sample yields observations $x_1, x_2, \ldots, x_{n_1}$?

8.4 Reconsider example 8.1 as a sequential prediction sampling inspection problem with the following changes.

$$N = 10, \quad k_1 = 0.2, \quad k_2 = 1.0, \quad l = 0.35,$$

(i) $\gamma = 0.40, \quad g = 1, \quad h = 1,$

(ii) $\gamma = 0.40, \quad g = 9, \quad h = 1,$

(iii) $\gamma = 0.30, \quad g = 1, \quad h = 1,$

(iv) $\gamma = 0.30, \quad g = 9, \quad h = 1.$

Compare your solutions with the optimum fixed size sample tests.

9
Regulation and optimisation

9.1 Introduction

In the next three chapters we focus attention on some predictive problems which occur in essentially regression-type situations. The distinctive features of regulation, optimisation, calibration and diagnosis have already been indicated in examples 1.4–1.7 in chapter 1 and are formally recognised in the classification of prediction problems in appendix II. In these regression-type situations we envisage a typical experiment f_t, performed at the particular value t of an 'independent' variable and resulting in the observation of a y-value. There is thus a class F of 'future' experiments:

$$F = \{f_t : t \in T\},$$

where each f_t has the same outcome space Y and T acts as index set for the class F. The informative experiment e consists of a single performance of each of the n independent experiments $f_{t_1}, f_{t_2}, \ldots, f_{t_n}$, and we shall denote the set of outcomes by $\mathbf{x} = (x_1, \ldots, x_n)$, and write $\mathbf{t} = (t_1, \ldots, t_n)$, $\mathbf{z} = \{(t_1, x_1), \ldots, (t_n, x_n)\}$ as in §2.5.

In a problem of control type we aim to obtain a specific y-value, $y = y_0$ say, and require to find a corresponding t-value which is optimum in some defined sense. In particular, in problems of regulation we know the value y_0 for which we are aiming and wish to regulate or control the t-value in an effort to attain this y_0. In optimisation problems we wish to choose the t-value in order to maximise (minimise) the y-value subject to constraints and are usually unaware of the optimal value. In problems of calibration and diagnosis we are given the y-value, and require some point or interval estimate for the corresponding t-value. The main difference between calibration and diagnosis on the one hand and regulation and optimisation on the other is that for the latter cases the choice of f_t is ours, whereas for calibration and diagnosis we have no control over which member of F is performed.

In this chapter we deal with regulation and optimisation problems. Since we are going to carry out a particular experiment f_t in the future (on the assumption that there is a unique optimal t-value), we are concerned only with point prediction. There is no advantage in providing an interval or set estimate, and indeed such a procedure may well lead to confusion.

The derivation of the appropriate predictive density function $p(y|t, \mathbf{z})$ performs the information-extraction process in the problem. As illustrated in §2.5 we can use table 2.3 to obtain these densities.

9.2 Point regulation

We again present the problem as one of statistical decision theory with the following components.

(i) *Parameter space*. Y plays the role of the set of unknown parameters, the unknown outcome of the future experiment f_t playing the role of an unknown state of nature. The predictive density $p(y|t, \mathbf{z})$ provides an assessment of the plausibility of a particular y.

(ii) *Action set*. The action set A is simply the set T, since we can choose to perform the future experiment at any $t \in T$.

(iii) *Utility function*. A sensible utility specification is one which for given t compares the y-value obtained with the optimal y_0. We thus obtain a utility structure $U(t, y, y_0)$ by defining a function on the product domain $A \times Y \times y_0$.

We require to maximise the expected utility

$$U(t, y_0) = \int_Y U(t, y, y_0)\, p(y|t, \mathbf{z})\, dy. \tag{9.1}$$

Since we have complete control over the choice of future experiment f_t we simply maximise $U(t, y_0)$ with respect to t to obtain the best member f_{t*} of F to perform. Thus

$$U(t^*, y_0) = \max_T U(t, y_0). \tag{9.2}$$

We now mention briefly a few simple utility functions $U(t, y, y_0)$ suitable for some regulation problems. Since the aim in regulation is to perform an experiment f_t which will give us a value of y very close to a specified y_0 we shall naturally attach penalties to values of y which differ from y_0.

All-or-nothing utility. If it is imperative that the y-value obtained should be very near y_0, the natural formulation for the utility is as the limiting case $(\epsilon \to 0)$ of utility function:

$$U(t, y, y_0) = \begin{cases} 1 & (y_0 - \epsilon \leqslant y \leqslant y_0 + \epsilon), \\ 0 & \text{otherwise.} \end{cases} \tag{9.3}$$

Then

$$\begin{aligned} U(t, y_0) &= \int_{y_0 - \epsilon}^{y_0 + \epsilon} p(y|t, \mathbf{z})\, dy \\ &= 2\epsilon\, p(y_0|t, \mathbf{z}) + o(\epsilon) \quad (\epsilon \to 0) \end{aligned} \tag{9.4}$$

provided certain simple regularity conditions on p are satisfied. We therefore need to maximise $p(y_0|t, z)$ with respect to t. This is intuitively reasonable and is analogous to maximum likelihood estimation. We are selecting the t-value which gives most support for y_0 in the sense that it produces the largest predictive probability of obtaining y_0.

Linear utility. Often a more realistic utility function is the piecewise linear type:

$$U(t, y, y_0) = \begin{cases} -\xi(y_0 - y) & (y < y_0), \\ -\eta(y - y_0) & (y \geqslant y_0), \end{cases} \tag{9.5}$$

where $\xi, \eta > 0$. We see that $U(t, y_0, y_0)$ is the maximum value. Then

$$\begin{aligned} U(t, y_0) &= -\xi \int_{-\infty}^{y_0} (y_0 - y) p(y|t, z) \mathrm{d}y \\ &\quad - \eta \int_{y_0}^{\infty} (y - y_0) p(y|t, z) \mathrm{d}y \\ &= -(\xi + \eta) \int_{-\infty}^{y_0} (y_0 - y) p(y|t, z) \mathrm{d}y \\ &\quad - \eta \{\mathrm{E}(y|t, z) - y_0\}. \end{aligned} \tag{9.6}$$

Our task is to select the value t^* of t which maximises $U(t, y_0)$. Unlike the direct case of linear-loss point prediction (§3.1) no simple interpretation, such as a quantile of the predictive density function, is available in the case of linear utility regulation.

Quadratic loss. Here

$$U(t, y, y_0) = -(y - y_0)^2. \tag{9.7}$$

Then

$$U(t, y_0) = -\mathrm{V}(y|t, z) - \{\mathrm{E}(y|t, z) - y_0\}^2, \tag{9.8}$$

and we must again choose the t^* value of t which maximises $U(t, y_0)$.

Cost function. A function $K(t)$ which takes into account the cost of performing the future experiment at value t, but which we assume to be independent of y, may easily be incorporated into the utility specification.

The method of subsequently maximising $U(t, y_0)$ must depend largely on its form and on the computing facilities available. In table 9.1 we list the optimum t-values in cases where simple formulations are available. In other cases, for example where the predictive density function is Student, iterative or search techniques are required, and we provide such an application in

Table 9.1 *Optimal t-values for some standard cases*

$p(y \mid t, z)$	Utility function All-or-nothing	Linear-loss	Quadratic-loss
$\mathrm{BeBi}(t, G, H)$	$\left[y_0 \frac{(G+H-1)}{G}\right]$	–	$\left[\frac{(G+H+1)(2y_0-1)}{2(G+1)}+1\right]$
$\mathrm{NeBi}\left(G, \frac{t}{t+H}\right)$	$\frac{H}{G} y_0$	Solution t^* of $I_{t^*/(t^*+H)}(y_0, G+1) = \frac{\xi}{\xi+\eta}$	$\frac{H(2y_0-1)}{2(G+1)}$
$\mathrm{InBe}(t, G, H)$	$\left[\frac{G}{H} y_0 + 1\right]$ (t integer)	–	$\frac{G-2}{H} y_0 - \frac{1}{2}$

example 9.2. First, however, we consider an illustrative example for one of the cases shown in table 9.1.

Example 9.1

Particle emission. The number of radioactive particles emitted in a unit time period by a sample of chemical compound depends on the amount t of radioactive material contained in the sample. Past experiments with prepared amounts $t_1, \dots, t_n$ of the radioactive material gave radioactive counts $x_1, \dots, x_n$. For future purposes we want a sample which will emit y_0 radioactive particles in the unit time period. What amount of radioactive material should be used in preparation of the sample?

Assuming the usual Poisson model for radioactive counts we may take f_t, the experiment which records the number of radioactive particles from a sample with amount t of the radioactive material, as $\text{Po}(t\theta)$. From sufficiency considerations we can clearly consider the informative experiment as recording $x = x_1 + \dots + x_n$ and then

$$p(x\,|\,\mathbf{t}, \theta) = \text{Po}(\Sigma t_i \theta)$$

and

$$p(y\,|\,t, \theta) = \text{Po}(t\theta).$$

Assuming a conjugate prior $p(\theta) = \text{Ga}(g, h)$ we immediately obtain from table 2.3 that

$$p(y\,|\,t, \mathbf{z}) = \text{NeBi}\left(G, \frac{t}{H+t}\right), \tag{9.9}$$

where $G = g + x, H = h + \Sigma t_i$.

If it is vital to obtain y_0, use of the all-or-nothing utility leads us to maximise

$$\frac{t^{y_0}}{(H+t)^{y_0+G}}$$

with respect to t. Hence

$$t^* = \frac{H}{G} y_0. \tag{9.10}$$

This is sensible when one recalls that $\text{E}(y\,|\,t, \mathbf{z}) = (G/H)\,t$. We are therefore choosing the t-value which makes the expected value of y equal to y_0.

Suppose that a linear utility function of form (9.5) is suitable. Then we find that the expected utility is maximised when t^* is the solution of

$$I_{t/(t+H)}(y_0, G+1) = \frac{\xi}{\xi + \eta}. \tag{9.11}$$

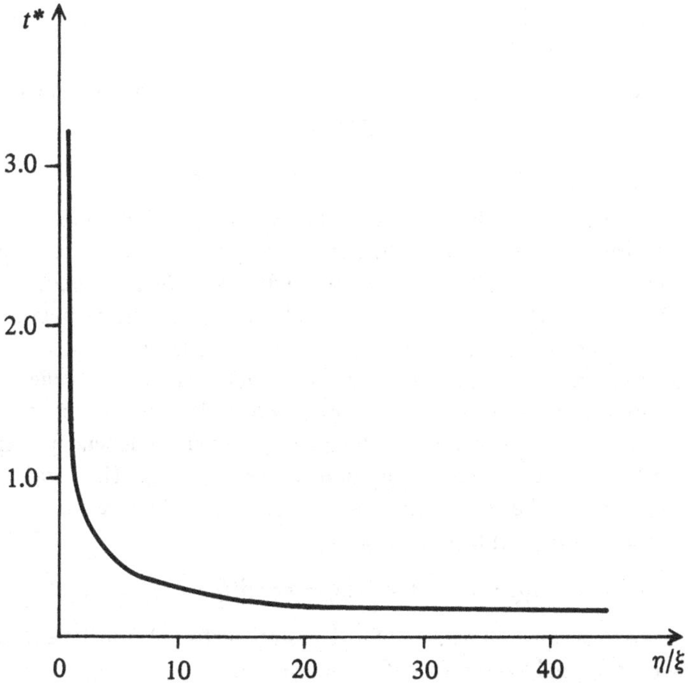

Fig. 9.1 Graph of t^* against η/ξ.

In the derivation use is made of relationship (A23) in appendix I between the negative binomial and beta distributions. As η/ξ increases we see that t^* decreases. This is because as η/ξ increases the penalty for obtaining a y-value greater than y_0 becomes relatively larger. We therefore aim for smaller y-values. Since on the average y increases as t increases ($\mathrm{E}(y \mid t, \theta) = t\theta$), our choice of t^* decreases. It is possible to display the dependence of t^* on η/ξ by plotting the graph $(\eta/\xi, t^*)$. To illustrate this point fig. 9.1 shows the graph for the case where $y_0 = 2, G = 25, H = 14$. When η/ξ is large (> 10 say) small variations in its value are of little importance. For smaller values of η/ξ, however, any variation is critical.

For a quadratic utility function,

$$\begin{aligned} U(t, y_0) &= -\mathrm{V}(y \mid t, \mathbf{z}) - \{\mathrm{E}(y \mid t, \mathbf{z}) - y_0\}^2 \\ &= \frac{-Gt(H+t)}{H^2} - \left\{\frac{Gt}{H} - y_0\right\}^2. \end{aligned} \tag{9.12}$$

Thus, provided $y_0 \neq 0$, we find that

$$t^* = \frac{H(2y_0 - 1)}{2(G+1)}. \tag{9.13}$$

We now consider an example where the maximisation problems are more involved and iterative or search techniques are required.

Example 9.2

Suppressor drug dose level. In a clinical trial of a new drug aimed at suppressing the level of a certain body hormone to a satisfactory level, 23 randomly chosen patients were each given a different dose of the drug and their subsequent hormone level recorded. We require to recommend a dose t of a drug which will suppress the level of the body hormone in a patient to a level y_0.

Suppose that we may take f_t, the experiment which records hormone level after application of dose t of the drug, to be described by a $\mathrm{No}(\alpha + \beta t, \tau)$ density. The informative experiment e then consists of the independent experiments $f_{t_1}, f_{t_2}, \ldots, f_{t_n}$ with recorded hormone levels $x_1, \ldots, x_n$. The parameters α and β occur only in the combination $\mu = \alpha + \beta t$ in f_t, and hence we may summarise the information from e by using

$$m = \bar{x} + \hat{\beta}(t - \bar{t}), \quad v = \Sigma\{x_i - \bar{x} - \hat{\beta}(t_i - \bar{t})\}^2,$$

where $\hat{\beta}$ is the least squares estimate of β. The statistic (m, v) is sufficient for (μ, τ) given $S(\mathbf{t}, \mathbf{t})$, and has a distribution of the form

$$p(m, v|\mu, \tau) = p(m|\mu, \tau)\, p(v|\tau),$$

$$p(m|\mu, \tau) = \mathrm{No}(\mu, k\tau), p(v|\tau) = \mathrm{Ch}(\nu, \tau), \tag{9.14}$$

where

$$\frac{1}{k} = \frac{1}{n} + \frac{(t-\bar{t})^2}{S(\mathbf{t}, \mathbf{t})}, \nu = n - 2.$$

We can therefore use table 2.3 in chapter 2 to update a prior $\mathrm{NoCh}(b, c, g, h)$ density for (μ, τ) and so obtain the predictive density function

$$p(y|t, \mathbf{z}) = \mathrm{St}\left\{G, B, \left(1 + \frac{1}{C}\right)\frac{H}{G}\right\} \tag{9.15}$$

where B, C, G, H are as defined in case 5 of table 2.3.

Suppose that any excess of hormone is assessed to be η times as harmful as a similar deficit. Then we may use

$$U(t, y, y_0) = \begin{cases} -(y_0 - y) & (y < y_0), \\ -\eta(y - y_0) & (y \geqslant y_0). \end{cases} \tag{9.16}$$

The expected utility is given by

$$U(t, y_0) = \left\{\left(1+\frac{1}{C}\right)\frac{H}{G}\right\}^{1/2}\left\{\eta w - (1+\eta)\frac{G+w^2}{G-1}\psi_G(w) - (1+\eta) w\Psi_G(w)\right\} \tag{9.17}$$

where

$$w = (y_0 - B)\left\{\left(1+\frac{1}{C}\right)\frac{H}{G}\right\}^{-1/2}$$

and ψ_G, Ψ_G are the probability density function and distribution function respectively of a t(G) random variable. In general no simple expression can be found for the t-value which maximises $U(t, y_0)$. Even the assumption of prior ignorance with $p(y \mid t, \mathbf{z}) = \mathrm{St}[\nu, m, \{1 + (1/k)\}v/\nu]$ does not simplify matters.

A similar problem exists if we consider the quadratic loss function (9.7). Then

$$U(t, y_0) = -\frac{(C+1)H}{C(G-2)} - (B - y_0)^2. \tag{9.18}$$

If we set the derivative with respect to t equal to zero then we have to find the solution of a polynomial equation of degree 5 in t. For the case of prior ignorance, however, a solution is readily obtained, and is given by

$$t^* = \bar{t} + \frac{(y_0 - \bar{x})}{\hat{\beta} + \dfrac{v}{\hat{\beta}(\nu-2)S(\mathbf{t}, \mathbf{t})}}. \tag{9.19}$$

The simplest classical solution would be to perform the future experiment at the value

$$t = \frac{y_0 - \hat{\alpha}}{\hat{\beta}} = \bar{t} + \frac{y_0 - \bar{x}}{\hat{\beta}}, \tag{9.20}$$

the result of solving the regression equation for t and substituting $y = y_0$. The use of such an estimate is, however, questionable since the variance of a predicted y-value using this t increases as t increases; in other words, the variance depends on the value y_0.

We now consider a specific numerical example. Suppose that the recommended hormone level y_0 is 65 and that any excess of hormone is twice as harmful as a similar deficit. We have available the results shown in table 9.2 from the informative experiment in which 23 patients received varying doses.

We have

$$y_0 = 65, \eta = 2, n = 23,$$

$$\bar{x} = 50.87, \bar{t} = 7.60,$$

$$\hat{\beta} = 12.14, S(\mathbf{t}, \mathbf{t}) = 26.11, v = 667.98.$$

For simplicity we assume prior ignorance, so that with utility function (9.16) the expected utility $U(t, y_0)$ reduces to

$$U(t, y_0) = 1.10\,(t^2 - 15.21t + 85.07)^{1/2} \times \{2w - (3.15 + 0.15w^2)\,\psi_{21}(w) - 3w\Psi_{21}(w)\} \quad (9.21)$$

where

$$w = \frac{96.42 - 11.00t}{(t^2 - 15.21t + 85.07)^{1/2}}.$$

A simple search technique reveals that $U(t, y_0)$ takes its maximum value when $t = 8.54$.

This may be compared with (i) the classical estimate (9.20) which gives $t = 8.77$ and (ii) the estimate (9.19) associated with the quadratic utility structure which gives $t = 8.76$.

9.3 Set regulation

Sometimes in regulation problems the objective may be not to obtain a particular value $y = y_0$ but to ensure that the y-value lies in some subset Y_0 of Y. Although we seek to regulate the outcome within a set we are still concerned with finding a single element t from T. Simple utility functions corresponding to those of the previous section are straightforwardly derived, but their use, even for the standard distributions, leads to formulations which are not as simple as those previously encountered. It seems that trial-and-error computations are necessary. We leave to the reader the simple task of formulating the optimisation problems and discovering the difficulties. We follow here the easier path of an illustrative example.

Example 9.3

Laminate design. Recall example 1.4 on laminate design discussed in chapter 1. If, for any particular sheet, we assume that the number of flaws is Poisson-distributed with mean θ, it follows immediately that the conditional distribution of the total number y of flaws given that t independently produced sheets have been superimposed is Poisson with mean $t\theta$, that is, $p(y\,|\,t, \theta)$ is $\mathrm{Po}(t\theta)$. A suitable utility function for this problem may take the following form:

$$U(t, y, y_0) = \begin{cases} K_1(t) & (y \leqslant y_0), \\ -K_2(t) & (y > y_0), \end{cases} \quad (9.22)$$

where $K_1(t)$, $K_2(t)$ are cost functions. If $y \leqslant y_0$ we shall be able to sell the product and thus make some profit $K_1(t)$, which we assume to depend only on the number t of sheets. If, however, $y > y_0$ the product will be rejected

Table 9.2 *Dosages of drug and body hormone levels for 23 patients*

Dose	t(ng/dl) :	9.4	7.8	6.8	8.5	5.9	6.4	8.8	8.0
Hormone level	x(ng/dl) :	71	65	35	62	26	43	61	52
Dose	:	8.1	6.1	8.2	6.1	7.0	7.5	8.7	9.3
Hormone level	:	53	28	60	35	42	47	71	69
Dose	:	8.1	7.2	8.3	8.4	5.8	7.9	6.6	
Hormone level	:	68	54	59	52	27	50	40	

and there will be a loss $K_2(t)$ associated with this, which we have again assumed to be a function of t. One could introduce some factor comparing y with the critical value y_0, but this increases the complexity of the maximisation of the expected utility. If the main interest is the acceptance or rejection of the finished product and not in the amount by which the number of faults falls short of, or exceeds, the critical value, then the above utility function seems the most satisfactory. For simplicity we take $K_1(t) = k_1 t$, $K_2(t) = k_2 t$, where k_1, k_2 are positive constants.

From table 2.3 we have that

$$p(y \mid t, z) \text{ is NeBi} \left(G, \frac{t}{t+H}\right),$$

where $G = g + x$, $H = h + \Sigma t_i$ as in (9.9). If we use (A23) of appendix I the expected utility may be written in the form

$$U(t, y_0) = k_1 t - (k_1 + k_2) t I_{t/(t+H)}(y_0 + 1, G), \tag{9.23}$$

and we select the t-value t^* which gives the largest value of this expected utility. For the numerical example given in example 1.4 with $k_1 = 10k_2$, we have, for prior ignorance on θ,

$$U(t, y_0) = k_2 t\{10 - 11 I_{t/(t+81)}(8, 49)\}.$$

Evaluating $U(t, y_0)/k_2$ for $t = 0, 1, 2, \ldots$ we obtain the solution $t^* = 9$, and it follows that the optimal policy is to produce an article by superimposing 9 sheets.

Similarly we may obtain optima for different values of the ratio k_1/k_2, and table 9.3 gives a selection of results.

Table 9.3 *Optimal number of sheets in laminate design problem*

k_1/k_2	Optimal number of sheets
1	8
2	8
3	9
4	9
5	9
10	9

9.4 Regulation problems with a finite index set

Certain types of classification problems may be considered under this statistical model. Suppose that there are k categories, indexed by the t-values $t = 1, 2, \dots, k$, to which an item may belong. In the future experiment f_t we may wish to obtain a certain y-value ($y = y_0$, say). For which category do we carry out the experiment? This is simply a regulation problem in which the index set T is not continuous, but finite. Again we consider only a simple example and leave the reader to derive solutions for other situations.

Suppose that we take $p(y\,|\,t, \boldsymbol{\theta}) = \text{No}(\mu_t, \tau)$, where $\boldsymbol{\theta} = (\mu_1, \mu_2, \dots, \mu_k, \tau)$, and assume that $p(\boldsymbol{\theta}) = \text{No}_k\text{Ch}(\mathbf{b}, \mathbf{c}, g, h)$, that is, $p(\boldsymbol{\mu}\,|\,\tau) = \text{No}_k(\mathbf{b}, \tau\mathbf{c})$, $p(\tau) = \text{Ch}(g, h)$. Let n_t observations in the informative experiment be in the tth category and suppose that m_t is the mean of the corresponding x-values ($t = 1, 2, \dots, k$). Then the predictive density $p(y\,|\,t, \mathbf{z})$ is

$$\text{St}\{G, B_t, (1 + C^{tt})H/G\} \tag{9.24}$$

where

$$\left.\begin{aligned}
\mathbf{B} &= (B_1, B_2, \dots, B_k)' = \mathbf{C}^{-1}(\mathbf{Dm} + \mathbf{cb}),\\
\mathbf{C} &= \mathbf{c} + \mathbf{D},\\
G &= g + n,\\
H &= h + \mathbf{x'x} + \mathbf{b'cb} - \mathbf{B'CB},\\
\mathbf{m} &= (m_1, m_2, \dots, m_k)',\\
\mathbf{D} &= \text{diag}\,(n_1, n_2, \dots, n_k),\\
n &= \sum_{t=1}^{k} n_t,\\
\mathbf{C}^{-1} &= \{C^{ij}\} \quad (i, j = 1, 2, \dots, k).
\end{aligned}\right\} \tag{9.25}$$

If we use utility function (9.5) then

$$\begin{aligned}
U(t, y_0) = \left((1 + C^{tt})\frac{H}{G}\right)^{1/2} \Bigg\{ &\eta w_t - (\xi + \eta)\frac{G + w_t^2}{G - 1}\,\psi_G(w_t)\\
&- (\xi + \eta)\, w_t\, \Psi_G(w_t)\Bigg\},
\end{aligned} \tag{9.26}$$

where

$$w_t = (y_0 - B_t)\left\{(1 + C^{tt})\frac{H}{G}\right\}^{-1/2} \quad (t = 1, 2, \dots, k).$$

We can carry out the maximisation simply by evaluating $U(t, y_0)$ for the k different values of t.

Table 9.4 *Lengths of 42 articles produced by 6 machines*

	Machine I	II	III	IV	V	VI
t	1	2	3	4	5	6
x (mm)	10.6	10.6	10.9	11.0	10.9	10.7
	11.2	11.0	11.2	10.9	11.1	10.9
	10.9	10.8	11.1	11.2	10.6	10.2
	11.3	10.8	11.2	10.9	10.9	10.2
	11.0	11.2	11.2	11.3	10.6	10.9
	11.0	10.4	11.3	11.3	10.7	10.7
	10.6		11.2	11.1		
			11.3	11.2		
			11.3			

Table 9.5 *Values of expected utility $U(t, y_0)$ (9.26) for the 6 machines*

t	n_t	$U(t, y_0)$
1	7	−0.19
2	6	−0.25
3	9	−0.24
4	8	−0.20
5	6	−0.25
6	6	−0.41

Example 9.4

Component length control. An item can be produced on any one of six machines I, II, ... , VI. The data in table 9.4 represent the lengths (mm) of a sample of items produced by the machines. If we require to produce an item of length 11.0 mm, which machine do we use?

If we assume prior ignorance on θ we have that

$$B_t = m_t, C^{tt} = 1/n_t, G = n, H = \sum_{i=1}^{n} x_i^2 - \sum_{t=1}^{k} n_t m_t^2 .$$

Taking $\xi = \eta = 1$ we find that the values of $U(t, y_0)$ given by (9.26) are as shown in table 9.5. The maximum value of $U(t, y_0)$ thus occurs when $t = 1$. We therefore perform the future experiment on Machine I.

9.5 Optimisation

In a number of problems of a regression nature we are interested in determining the value of the independent variable t at which we should perform a future experiment in order to obtain an optimal value of the dependent variable y. In some cases (for instance, certain curvilinear regression problems) the optimal value may simply be the maximum (minimum) value of y we can hope to obtain. In other cases the cost of performing the experiment f_t may vary with t and we shall then want to balance this cost against the benefit from a large (small) y-value. For optimisation problems, in contrast to regulation problems, we do not know the specific y_0 we are trying to obtain. Our objective is to optimise the outcome y of f_t subject to certain restrictions.

In the classification of prediction problems (appendix II) the components of the optimisation problem are as follows. The future experiment is one of the class $F = \{f_t : t \in T\}$; the predictive density function is of the form $p(y|t, \mathbf{z})$; the action set A is T; and the domain of the utility function is $A \times Y$. The only difference from regulation problems is therefore the change in the specification of the utility function, brought about by our lack of knowledge of y_0.

A suitable specification for the cases mentioned above takes the form

$$U(t,y) = y - K(t), \tag{9.27}$$

where $K(t)$ is a function which takes account of the cost of performing the future experiment f_t, but which is assumed to be independent of y. The expected utility is then

$$\begin{aligned} U(t) &= \int_Y U(t,y)p(y|t,\mathbf{z})\mathrm{d}y \\ &= \mathrm{E}(y|t,\mathbf{z}) - K(t). \end{aligned} \tag{9.28}$$

Obviously the solution to the optimisation problem depends critically on $K(t)$ and the form of T. For the standard distributions of cases 1, 2 and 3 in table 2.3, the expected values $\mathrm{E}(y|t, \mathbf{z})$ are linear in t. Hence use of a linear cost function $K(t)$ would result in the optimum t value being one of the extreme values of T.

If we are concerned with maximising a profit we may extend the above specification to the form

$$U(t,y) = f(y) - K(t), \tag{9.29}$$

where $f(y)$ represents the return or gain from an experiment with outcome y.

In these optimisation problems we are trying to determine the conditions (the value of the independent variable t, which may be vector-valued) under which we should perform the future experiment in order to obtain the optimum result. The classical approach involves response surfaces and their polynomial representation. Typically, to find a maximum response, the response surface is first estimated, for example by the method of least squares, and then the optimum point found by differentiation. Techniques such as the method of steepest ascent seek an approach to a stationary point, possibly by some sequential method of experimentation. We are here concerned more with the terminal rather than the preposterior analysis, however, and so we are not concerned with the design of the informative experiment e.

Consider first the case where there is an independent variable t which is real-valued. If we can assume that we are dealing with normal variability from the response surface then we can take

$$p(y\,|\,t,\boldsymbol{\theta}) = \text{No}(\phi(t,\boldsymbol{\beta}),\tau),$$

where $\boldsymbol{\theta} = (\boldsymbol{\beta},\tau)$ and $\phi(t,\boldsymbol{\beta})$ is the response surface. Thus, for polynomial regression, we have that

$$\phi(t,\boldsymbol{\beta}) = \beta_0 + \beta_1 t + \beta_2 t^2 + \dots + \beta_r t^r,$$

where $\boldsymbol{\beta} = (\beta_0, \beta_1, \dots, \beta_r)'$. The informative experiment consists of performances of $f_{t_1}, f_{t_2}, \dots, f_{t_n}$ yielding observations $x_1, x_2, \dots, x_n$.

By a simple extension of the arguments in §2.5 we can again make use of case 5 in table 2.3. For, if

$$\mathbf{T} = \begin{pmatrix} 1 & t_1 & t_1^{\,2} \dots t_1^{\,r} \\ 1 & t_2 & t_2^{\,2} \dots t_2^{\,r} \\ & \cdot & \\ & \cdot & \\ & \cdot & \\ 1 & t_n & t_n^{\,2} \dots t_n^{\,r} \end{pmatrix}, \quad \mathbf{x} = \begin{pmatrix} x_1 \\ x_2 \\ \cdot \\ \cdot \\ \cdot \\ x_n \end{pmatrix}$$

$$\mathbf{T}_0 = (1 \quad t \quad t^2 \dots t^r),$$

and if $\hat{\boldsymbol{\beta}} = (\mathbf{T}'\mathbf{T})^{-1}\mathbf{T}'\mathbf{x}$, the least-squares estimate of $\boldsymbol{\beta}$, then $m = \mathbf{T}_0\hat{\boldsymbol{\beta}}$ and $v = \mathbf{x}'\mathbf{x} - \mathbf{x}'\mathbf{T}\hat{\boldsymbol{\beta}}$ are jointly sufficient for $\mu = \phi(t,\boldsymbol{\beta})$ and τ. Furthermore m and v are independently distributed with

$$\left.\begin{aligned} p(m\,|\,\mu,\tau) &= \text{No}[\mu, \{\mathbf{T}_0(\mathbf{T}'\mathbf{T})^{-1}\mathbf{T}_0'\}^{-1}\tau], \\ p(v\,|\,\tau) &= \text{Ch}(n-r-1,\tau). \end{aligned}\right\} \tag{9.30}$$

The conjugate prior distribution for (μ,τ) is taken to be NoCh(b,c,g,h). Hence we have that

$$p(y\,|\,t,\mathbf{z}) = \text{St}\left\{G, B, \left(1+\frac{1}{C}\right)\frac{H}{G}\right\} \tag{9.31}$$

where B, C, G, H are as given in case 5 of table 2.3 with

$$k = (\mathbf{T}_0(\mathbf{T}'\mathbf{T})^{-1}\mathbf{T}_0')^{-1}, K = 1, \nu = n-r-1.$$

With utility function (9.27) we then have that

$$U(t) = B - K(t).$$

If we assume prior ignorance on (μ,τ) that is $c \to 0, g \to 0, h \to 0$, and let $K(t)$ be constant over the region of interest we see that $B = m = \mathbf{T}_0(\mathbf{T}'\mathbf{T})^{-1}\mathbf{T}'\mathbf{x}$.

Table 9.6 *Responses with varying amounts of drug for 34 patients*

Drug	Response	Drug	Response	Drug	Response
0.2	1.8	0.6	3.3	1.5	5.0
0.3	1.1	0.7	3.5	1.6	6.6
0.4	3.0	0.9	6.4	1.7	5.7
0.4	1.3	1.0	5.8	1.8	5.5
0.4	3.2	1.2	6.0	1.9	5.9
0.5	4.9	1.3	5.0	2.0	5.2
0.5	4.4	1.4	6.1	2.1	4.0
0.5	2.7	1.4	5.5	2.2	4.7
0.5	2.1	1.4	5.0	2.3	2.7
0.6	3.1	1.4	6.6	2.4	3.5
0.6	3.4	1.4	6.0	2.6	2.3
		1.4	5.9		

Then the optimum t^* at which we should perform the future experiment corresponds exactly to the maximum of the fitted least-squares regression curve and would be the optimum obtained by classical methods.

Example 9.5

Drug response. In a series of medical experiments 34 patients suffering from a certain complaint were treated with varying amounts of a drug and their responses recorded. It is known that underdoses and overdoses generally lead to smaller values of the response. The results of the series of trials of the experiment are shown in table 9.6, where x denotes the response measured when an amount t ml of the drug was used. One factor of interest would be to find the amount of the drug needed to maximise the response for a future patient.

Suppose that we assume a quadratic response surface, so that $r = 2$. For prior ignorance on $\boldsymbol{\theta}$ we find that

$$\hat{\boldsymbol{\beta}} = \begin{bmatrix} -0.40 \\ 8.65 \\ -2.98 \end{bmatrix}.$$

Hence

$$U(t) = -0.40 + 8.65t - 2.98t^2 - K(t).$$

If $K(t)$ is a constant over the region of interest then

$$t^* = 1.45.$$

The cost of an individual experiment may consist of two parts – the cost of the amount t of the drug and the charge for performing the treatment. We could then take $K(t) = k_1 + k_2 t$ where k_1, k_2 are positive constants. In this case we should perform the future experiment with an amount t of drug given by

$$t^* = \frac{8.65 - k_2}{5.96}$$

provided that $k_2 < 8.65$.

We may extend the ideas to the case where the independent variable t is vector-valued. For example, for two independent variables t_1 and t_2 a polynomial representation of the response surface is

$$\phi(t_1, t_2, \boldsymbol{\beta}) = \beta_0 + \beta_1 t_1 + \beta_2 t_2 + \beta_{11} t_1^2 + \beta_{12} t_1 t_2 + \beta_{22} t_2^2 + \ldots .$$

The predictive density function is as before except that

$$\mathbf{T} = \begin{bmatrix} 1 & t_{11} & t_{21} & t_{11}^2 & t_{11}t_{21} & t_{21}^2 & \ldots \\ 1 & t_{12} & t_{22} & t_{12}^2 & t_{12}t_{22} & t_{22}^2 & \ldots \\ \cdot \\ \cdot \\ \cdot \\ 1 & t_{1n} & t_{2n} & t_{1n}^2 & t_{1n}t_{2n} & t_{2n}^2 & \ldots \end{bmatrix},$$

and $\mathbf{T}_0 = [1 \quad t_1 \quad t_2 \quad t_1^2 \quad t_1 t_2 \quad t_2^2 \ldots]$.

Computations are considerably eased if e is well-designed and the t_1, t_2 values scaled.

Example 9.6

Maximising the yield of an industrial process. Recall example 1.5. In a balanced informative experiment the yields (kg), recorded in table 1.4, were obtained when an industrial process was run successively at five different temperatures and three different pressures, each combination of temperature and pressure being used twice. For simplification of the calculations we scale the values of temperature to $-2, -1, 0, 1, 2$ and of pressure to $-1, 0, 1$, so that $\Sigma t_{1i} = \Sigma t_{2i} = 0$. Also we take the response surface to be of second degree and given by

$$\phi(t_1, t_2, \boldsymbol{\beta}) = \beta_0 + \beta_1 t_1 + \beta_2 t_2 + \beta_{11}(t_1^2 - 2) + \beta_{12} t_1 t_2 + \beta_{22}(t_2^2 - 3/2).$$

This slight alteration in the response curve, and the corresponding change in $\mathbf{T}$ ensure that for the balanced design given $\mathbf{T}'\mathbf{T}$ is diagonal, and so we have greatly simplified the problem of inverting a 6×6 matrix.

If we again assume prior ignorance on $\boldsymbol{\theta}$ we have that

$$\hat{\boldsymbol{\beta}} = (\mathbf{T}'\mathbf{T})^{-1}\mathbf{T}'\mathbf{x} = \begin{bmatrix} 75.27 \\ 1.53 \\ 1.60 \\ -1.00 \\ -1.10 \\ -3.20 \end{bmatrix},$$

so that

$$U(t_1, t_2) = 75.27 + 1.53t_1 + 1.60t_2 - 1.00(t_1{}^2 - 2)$$
$$- 1.10\, t_1 t_2 - 3.20(t_2{}^2 - 3/2) - K(t).$$

Hence for a constant cost function the optimal operational values are

$$t_1{}^* = 0.69, \quad t_2{}^* = 0.13,$$

that is at 76.9°C and 1.28 atmospheres.

History

The Bayesian decision theory models derived in the chapter are given in Dunsmore (1969). Zellner and Chetty (1965) consider a similar regulation problem with a quadratic loss function for the multiple regression model. Lindley (1968) also considers this setting with the additional option of choosing which independent variables should be used in the attempt to control the y-value. His loss function contains a factor which depends on the values of the independent variables used.

The classical approach to optimisation is through the field of response surfaces; see, for example, Davies (1960).

Similar problems arise in the theory of stochastic control. There is considerable literature in this field and much Bayesian decision theory work; see, for example, Aoki (1967) and Sawagari, Sunahara and Nakamizo (1967).

Problems

9.1 A faculty of a state university is faced with the problem of the number of applicants to whom it should offer places. Study of the acceptance rate θ in previous years has shown a variation well described by a Be(G, H) distribution. The optimum number of students in first year is reckoned to be y_0. The disutilities of missing this target are estimated at ξ for each student place unfilled and η for each student in excess of the target. How many students should be offered places?

9.2 Complete the analysis of problem 1.4.

9.3 The rate of flow of air through doorways in a certain type of hospital area has been found to be approximately linearly related to door area for fixed temperature difference between the rooms separated by the door. For the temperature drop proposed between two rooms the data pairs (area, airflow) available are $(t_1, x_1), \ldots, (t_n, x_n)$. From considerations of comfort it has been agreed that the desirable airflow rate is y_0, and that any rate of airflow in excess of y_0 is three times as uncomfortable as the rate of airflow which falls a corresponding amount short of y_0. What door area should be adopted? With this adoption for what proportion of time is the airflow likely to exceed y_0?

9.4 The crushing strength of mortar varies at different times after mixing. For mortar prepared from two different types of cement A, B the crushing strengths are observed at six different times after mixing, four replications of each observation being made with the following results (in kg/cm^2).

Cement	Days, t: 2	4	8	16	32	64
A	320	350	420	480	560	610
	280	390	460	520	550	580
	300	330	450	530	540	600
	310	350	480	550	550	640
B	270	310	450	510	600	670
	220	340	460	530	610	700
	250	350	490	550	590	670
	220	310	430	560	560	710

For each of the mortar mixes determine after how many days you could continue with construction if a crushing strength of 400 kg/cm^2 is required. If you use utility function (9.5) what happens if $\xi \gg \eta$?

Which cement would you use if you require a crushing strength of (at least) 400 kg/cm^2 after 6 days?

9.5 For the preparation of a new household insecticide a manufacturer wishes to determine the quantity t of the highly potent active ingredient which he should include. He carries out a series of trials with various levels $t_1, t_2, \ldots, t_n$ of this ingredient and records the numbers of insects killed as shown below. What minimum level of the active ingredient should he use in his insecticide if he requires a 75 per cent success rate?

Dose (μg) :	0.1	0.2	0.3	0.4	0.5	0.6	0.7	0.8	0.9	1.0	1.1	1.2
No. submitted :	100	120	95	103	111	93	101	100	105	97	103	103
No. killed :	6	26	46	54	73	62	71	75	89	82	90	94

Suppose further that a utility structure of the form (9.5) is suitable with $\xi = 2, \eta = 1$ and with a cost component $k \log t$. What level would you now suggest?

10
Calibration

10.1 The nature of a calibration problem

Calibration is commonly regarded as the process whereby the scale of a measuring instrument is determined or adjusted on the basis of an informative or 'calibration' experiment. For example, if we wish to calibrate an unscaled thermometer we might note the position x_1 on the liquid scale when the thermometer is immersed in boiling water at atmospheric pressure, that is, corresponding to temperature t_1 (= 100°C); and the position x_2 when the immersion is in ice, say corresponding to temperature t_2 (= 0°C). We might then divide the scale between x_1 and x_2 into 100 equal divisions so that, when the thermometer is immersed into some other substance, we are able to deduce very simply from the x-scale the corresponding temperature of the substance. In this example the use of the calibration experiment yielding trial records (t_1, x_1), (t_2, x_2) is straightforward since there is, or at least we are assuming that there is, a one-to-one correspondence between the x-scale and the temperature or t-scale. But the same type of problem arises commonly in a less simple form, for usually, as in the following examples, there is no unique x corresponding to a given t.

Example 10.1

Measuring water content of soil specimens. Two methods are available for obtaining the water content in soil specimens. The first method, performed in the laboratory, is very accurate but is expensive and tedious to operate. The second method, which can be performed on site, is much quicker and cheaper, but is less accurate. It is intended that for future samples the second method be used and from the value obtained some estimate be made of the reading which the accurate first method would have given. Information on the relative values given by the two methods is obtained from a calibration experiment in which the water contents of 16 naturally occurring soil specimens were measured with results as shown in table 10.1.

The data $\mathbf{z}$ from the calibration experiment consist of the 16 paired measurements

$$\mathbf{z} = \{(t_i, x_i): i = 1, ..., 16\},$$

Table 10.1 *Water contents (percentages by weight) of 16 soil specimens determined by two methods*

Serial no. of specimen	Laboratory method	On-site method
1	35.3	23.7
2	27.6	20.2
3	36.2	24.5
4	21.6	15.8
5	39.8	29.2
6	24.1	17.8
7	16.1	10.1
8	27.5	19.0
9	33.1	24.3
10	12.8	10.6
11	23.1	15.2
12	19.6	11.4
13	26.1	19.7
14	19.3	12.7
15	18.8	12.6
16	39.8	31.8

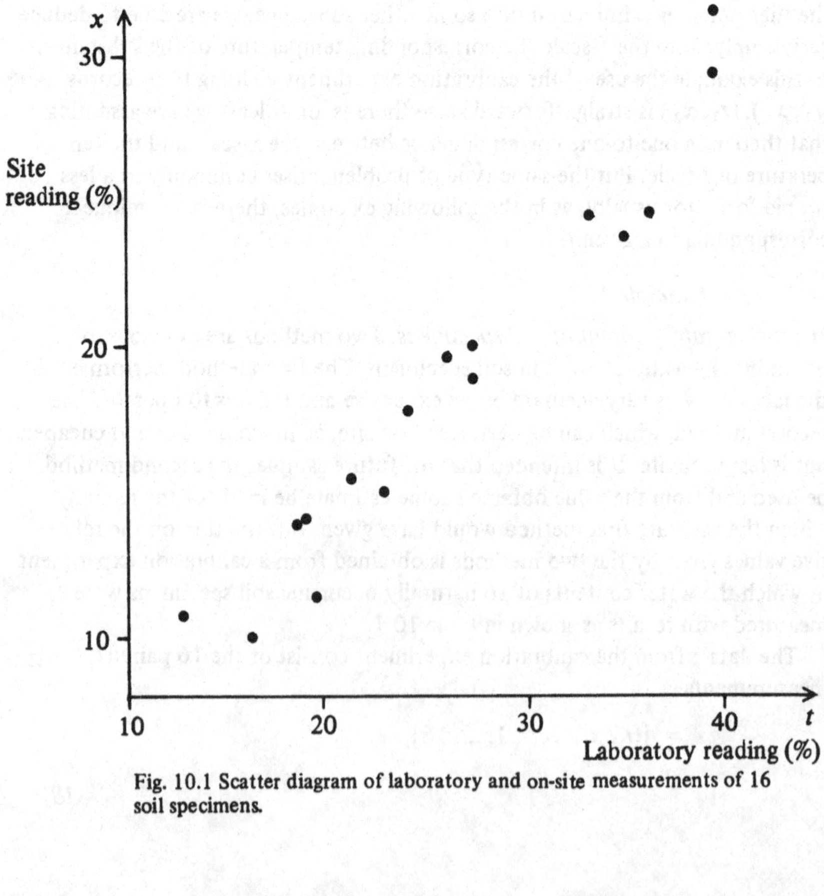

Fig. 10.1 Scatter diagram of laboratory and on-site measurements of 16 soil specimens.

where t_i and x_i denote measurements by the laboratory and on-site methods respectively. The associated scatter diagram of fig. 10.1 shows that we can no longer make the simplifying assumption that there is a unique x-value corresponding to a given t-value. For example $t_5 = t_{16} = 39.8$, but $x_5 \neq x_{16}$. To construct an appropriate calibration model we must therefore regard the on-site determination of water content corresponding to a true water content t determined by the laboratory method, as a random experiment f_t, say, with outcome space X. The calibration experiment then consists of recording

$$\mathbf{t} = (t_1, ..., t_{16}),$$

the vector of true values and the vector

$$\mathbf{x} = (x_1, ..., x_{16}),$$

where $x_1, ..., x_{16}$ are the outcomes of independent performances of $f_{t_1}, ..., f_{t_{16}}$. Moreover, for a new soil specimen with on-site measurement y the corresponding true value u is unknown, and so we are forced to consider the class F of experiments:

$$F = \{f_t : t \in T\},$$

where T is the set of possible true water contents.

We have thus a regression-type framework similar to that used for the regulation and optimisation problems of chapter 9. The difference is that whereas in the previous problems we had to select some suitable future experiment f_t, here the 'future' experiment f_u has already been performed and our objective is to attempt to identify the index u from which the known outcome y has arisen. A calibration problem is thus a kind of inverse prediction, a problem of retrospection, but we shall see shortly that in its resolution the predictive distribution plays a central role.

Example 10.2

Calibration of an autoanalyser. It is proposed to install a new autoanalyser in a hospital for the routine determination of the concentration of a certain enzyme in blood plasma samples. The enzyme concentration can be determined accurately by a long and costly laboratory method whereas the autoanalyser method is quick and cheap. It is known from the considerable past experience that the samples presented for analysis have enzyme concentrations (meq/l) which are normally distributed with mean 4.6 and standard deviation 0.8.

To evaluate the effectiveness of the autoanalyser 9 plasma samples, selected to cover the range of enzyme concentrations, have each been divided into four aliquots, one aliquot being assigned to the laboratory method and the other three to separate analyser determinations. The results are shown

Table 10.2 *Blood plasma concentrations of an enzyme determined by two methods*

Serial no. of plasma sample	Laboratory determination (meq/l)	Corresponding autoanalyser determinations (meq/l)
1	3.0	2.3, 2.4, 2.5
2	3.4	2.6, 2.8, 2.8
3	3.8	3.0, 3.0, 3.1
4	4.2	3.2, 3.3, 3.4
5	4.6	3.7, 3.7, 3.7
6	5.0	3.9, 4.0, 4.1
7	5.4	4.2, 4.2, 4.3
8	5.8	4.6, 4.7, 4.8
9	6.2	4.9, 5.0, 5.2

in table 10.2. In the future it is hoped that only one such aliquot need be analysed by the autoanalyser to provide a reliable estimate of enzyme concentration. Is this a reasonable hope?

For an aliquot from a new plasma sample the autoanalyser gives a reading of 3.8 meq/l. What can be said about enzyme concentration?

Fig. 10.2 shows the (t, x) scatter diagram for the data from this calibration experiment. As appears reasonable from this diagram, we make the assumption that all autoanalyser determinations are statistically independent even though they may be associated with aliquots from the same plasma sample. Thus the data from this informative experiment constitute

$$\mathbf{z} = \{(t_i, x_i): i = 1, \ldots, 27\},$$

with $(t_1, x_1) = (3.0, 2.3)$, $(t_2, x_2) = (3.0, 2.4)$, ..., $(t_{27}, x_{27}) = (6.2, 5.2)$. These are trial records of independent performances of $f_{t_1}, \ldots, f_{t_{27}}$, where f_t denotes the experiment which records the autoanalyser determination associated with an enzyme concentration t determined by the laboratory method.

The important difference between example 10.1 and the present example lies in the manner in which the t-values $t_1, \ldots, t_n$ in the calibration experiment have arisen. In the former, the t-values were naturally occurring and if it is to be assumed that future soil specimens arise in the same way as in the past then the calibration experiment provides some indication of what this pattern is. In the latter, the blood samples used in the informative experiment have been deliberately selected to give a reasonable cover of the set T, and improve the 'design' of the calibration experiment. Thus while their choice presumably reflects some view as to the future pattern of t-values the calibration experiment itself provides no new information about that pattern.

To distinguish clearly between these two types of calibration experiment we shall use the terms *natural* and *designed*.

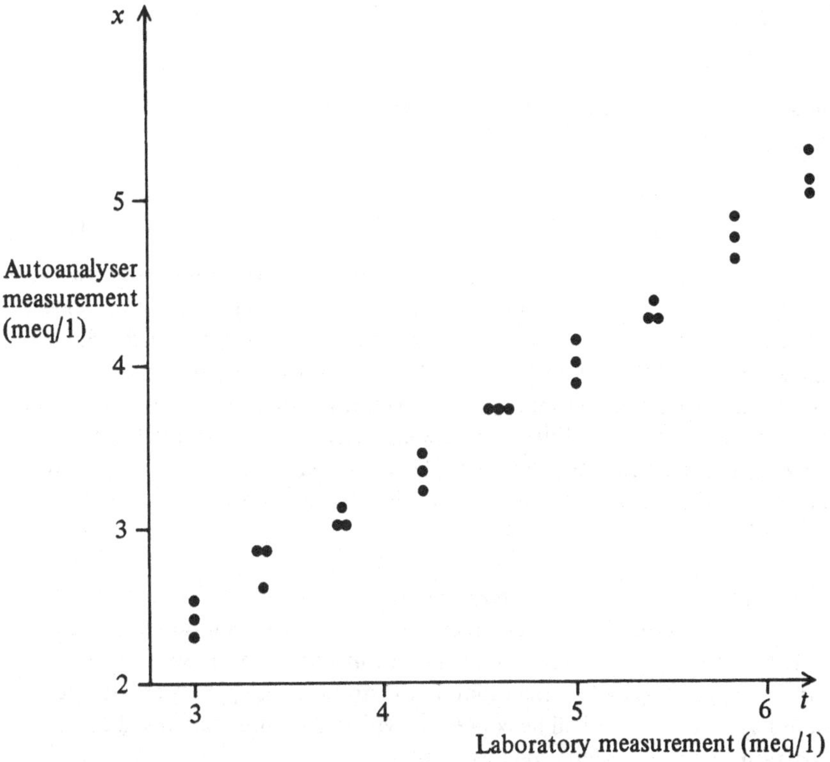

Fig. 10.2 Scatter diagram of laboratory and autoanalyser measurements of 9 plasma samples.

10.2 The calibrative distribution

The calibration problem can now be stated in general terms. Let us denote by the term *subject* any candidate for calibration such as the soil specimens and blood samples of the examples in § 10.1. Associated with each subject is a unique element of a specified set T and we shall refer to this element as the *index* of the subject. In our two examples the indices are the accurate laboratory determinations of water content and enzyme concentration. The aim of calibration is to identify the indices of subjects as clearly or as closely as possible. For each subject it is possible to perform a *trial* yielding an outcome or *measurement* in a specified record set X. Such a trial is an experiment belonging to a class

$$F = \{f_t : t \in T\}$$

of experiments. The trial associated with a subject whose index is t is f_t. It is assumed that for a number n of subjects it has been possible to record both the index and the corresponding measurement. The n *trial records*

$(t_1, x_1), \ldots, (t_n, x_n)$ thus constitute a calibration experiment with data

$$\mathbf{z} = \{(t_1, x_1), \ldots, (t_n, x_n)\}, \tag{10.1}$$

which we may also conveniently write in the form

$$\mathbf{z} = (\mathbf{t}, \mathbf{x}), \tag{10.2}$$

where

$$\mathbf{t} = (t_1, \ldots, t_n), \quad \mathbf{x} = (x_1, \ldots, x_n). \tag{10.3}$$

A new subject of unknown index u is presented and when the associated trial is performed the measurement obtained is y. In the light of our knowledge of this measurement y and of the data $\mathbf{z}$ of the calibration experiment what can we say about the unknown index u? Clearly this will depend on any assumptions we make concerning the probabilistic mechanism by which trial records are generated, and these we now examine. Our objective is to try to arrive at a probabilistic description $p(u \mid y, \mathbf{z})$, expressing the plausibility of the various u for the given values of y and $\mathbf{z}$. This distribution we shall refer to as the *calibrative distribution*.

Calibrative distribution for a natural calibration experiment. We now set out carefully the assumptions under which the calibrative distribution is derived. Only by exposing the assumptions can we be in a position to examine whether for a particular application the calibrative distribution is appropriate. In each of our applications we shall be concerned with parametric models and so this constitutes our first calibration assumption.

C1
The class of probability models describing the generation of a trial record is parametric.

Denote the parameter set by Ω, so that the density function corresponding to parameter $\omega \in \Omega$ is $p(t, x \mid \omega)$. The next two assumptions concern the nature of the parameter ω. We envisage ω as having two components $\psi \in \Psi$ and $\theta \in \Theta$ and the assumptions C2 and C3 give meanings to these two aspects of the parameter.

C2
$p(t \mid \psi, \theta) = p(t \mid \psi)$.

Here we are simply stating that ψ is the parameter which is concerned with the natural arrival pattern of the indices or t-values. Given ψ, knowledge of θ in no way affects the arrival pattern. We shall thus term ψ the *arrival* parameter.

C3
$p(x \mid t, \psi, \theta) = p(x \mid t, \theta)$.

This assumption concerns the description of the trial f_t and asserts that its description involves only the θ component of the parameter. The parameter θ is therefore important in describing the structure of the regression experiments in the class F, and so we shall term it the *structural* parameter.

Assumptions C1–3 then combine to give us the description of the variability in trial records:

$$p(t, x \mid \omega) = p(t, x \mid \psi, \theta)$$
$$= p(t \mid \psi) p(x \mid t, \theta). \qquad (10.4)$$

Another simplifying assumption is that subjects or trial records are statistically independent. This is certainly an assumption that may require very careful scrutiny for each application. In § 10.1 we have already examined its relevance for example 10.2.

C4
For any set $\mathbf{z}$ of trial records $z_1, z_2, \ldots$

$$p(\mathbf{z} \mid \omega) = \Pi p(z_i \mid \omega).$$

We now make an assumption about our state of knowledge about the arrival parameter ψ and the structural parameter θ prior to the calibration experiment. This states that our initial sources of information are independent.

C5

$$p(\psi, \theta) = p(\psi) p(\theta).$$

Finally, since we are dealing with the case of a natural calibration experiment we assume that further subjects presenting will follow the same pattern of arrival as the original trial subjects.

C6

$$p(u \mid \psi, \theta) = p(u \mid \psi),$$

which is of the same form as the $p(t \mid \psi)$ of C2.

In following through the consequences of these assumptions we must first recognise that we have three unknown indices or parameters, u, ψ and θ, and that the set of data is $y, \mathbf{z}$. The technique for obtaining $p(u \mid y, \mathbf{z})$ is then simply to use the assumptions and Bayes's theorem to obtain first $p(u, \psi, \theta \mid y, \mathbf{z})$ and then to integrate out ψ, θ to derive the marginal $p(u \mid y, \mathbf{z})$ as the calibrative distribution. First we obtain the forms of the prior $p(u, \psi, \theta)$ and of the likelihood $p(y, \mathbf{z} \mid u, \psi, \theta)$ as consequences of the assumptions. We have

$$p(u, \psi, \theta) = p(u \mid \psi, \theta) p(\psi, \theta)$$
$$= p(u \mid \psi) p(\psi) p(\theta) \qquad (10.5)$$

by C5 and C6.

Also since (u, y) is a trial record like $(t_1, x_1), \ldots, (t_n, x_n)$ we have

$$\begin{aligned} p(u,y;\mathbf{z}|\psi,\theta) &= p(u,y|\psi,\theta)\prod_{i=1}^{n} p(z_i|\psi,\theta) \text{ by C4} \\ &= p(u|\psi)p(y|u,\theta)\prod_{i=1}^{n} p(t_i|\psi)\prod_{i=1}^{n} p(x_i|t_i,\theta) \\ &\qquad\qquad\qquad\qquad\qquad\qquad \text{by C1–3} \\ &= p(u|\psi)p(y|u,\theta)p(\mathbf{t}|\psi)p(\mathbf{x}|\mathbf{t},\theta) \qquad (10.6) \end{aligned}$$

in an obvious shortened notation. Hence, from C6,

$$\begin{aligned} p(y,\mathbf{z}|u,\psi,\theta) &= \frac{p(u,y;\mathbf{z}|\psi,\theta)}{p(u|\psi)} \\ &= p(y|u,\theta)p(\mathbf{t}|\psi)p(\mathbf{x}|\mathbf{t},\theta) \qquad (10.7) \end{aligned}$$

by (10.6). Now applying Bayes's theorem and using (10.5) and (10.7) we have

$$\begin{aligned} p(u,\psi,\theta|y,\mathbf{z}) &\propto p(u,\psi,\theta)p(y,\mathbf{z}|u,\psi,\theta) \\ &\propto p(u|\psi)p(\psi)p(\mathbf{t}|\psi)p(\theta)p(\mathbf{x}|\mathbf{t},\theta)p(y|u,\theta) \\ &\propto p(u|\psi)p(\psi|\mathbf{t})p(\theta|\mathbf{z})p(y|u,\theta), \qquad (10.8) \end{aligned}$$

where

$$p(\psi|\mathbf{t}) \propto p(\psi)p(\mathbf{t}|\psi), \qquad (10.9)$$

$$p(\theta|\mathbf{z}) \propto p(\theta)p(\mathbf{x}|\mathbf{t},\theta) \qquad (10.10)$$

are the post-calibration experiment probability assessments for ψ and θ. Now integrating out ψ and θ we have the calibrative distribution:

$$p(u|y,\mathbf{z}) \propto p(u|\mathbf{t})p(y|u,\mathbf{z}), \qquad (10.11)$$

where

$$p(u|\mathbf{t}) = \int_{\Psi} p(u|\psi)p(\psi|\mathbf{t})d\psi, \qquad (10.12)$$

$$p(y|u,\mathbf{z}) = \int_{\Theta} p(y|u,\theta)p(\theta|\mathbf{z})d\theta. \qquad (10.13)$$

We can now see the relevance of the predictive distribution to the calibration problem, for (10.13) is simply the predictive distribution associated with the 'future' experiment f_u and based on data $\mathbf{z}$. Indeed (10.11) takes the form of Bayes's theorem, as it could be applied after the data $\mathbf{z}$ or $(\mathbf{t}, \mathbf{x})$ of the calibration experiment are known. The prior plausibility assessment is then $p(u|\mathbf{t})$ based on the pattern we have seen in $t_1, \ldots, t_n$, and the 'likelihood' of Bayes's theorem takes the form of the predictive distribution.

Calibrative distribution associated with a designed calibration experiment. When the t-values $t_1, \dots, t_n$ for the calibration experiment are selected by the experimenter then the calibration experiment provides no direct information concerning the plausibilities of various u. We no longer require to describe how trial records $(t_1, x_1), \dots, (t_n, x_n)$ are generated but only the nature of the given regression experiments $f_{t_1}, \dots, f_{t_n}$. For this we have the independence assumption:

C′1

$$p(\mathbf{x}\,|\,\mathbf{t}, \theta) = \Pi p(x_i\,|\,t_i, \theta).$$

The second assumption is concerned with a typical new subject with trial record (u, y):

C′2

$$p(u, y\,|\,\theta) = p(u)p(y\,|\,u, \theta).$$

This incorporates the assumption that the calibration experiment provides no information about u, since it implies that $p(u\,|\,\theta) = p(u)$. If, however, we select the t-values in the informative experiment we clearly have some view as to what indices are likely to turn up, and the onus is on us to describe this view directly in $p(u)$. The calibrative distribution can then be arrived at by the following sequence of steps:

$$\begin{aligned} p(u, \theta\,|\,y; \mathbf{t}, \mathbf{x}) &\propto p(u, \theta)p(y; \mathbf{t}, \mathbf{x}\,|\,u, \theta) \\ &\propto p(u)p(\theta)p(y\,|\,u, \theta)p(\mathbf{x}\,|\,\mathbf{t}, \theta) \\ &\propto p(u)p(y\,|\,u, \theta)p(\theta\,|\,\mathbf{z}) \end{aligned} \qquad (10.14)$$

and integrating out θ,

$$p(u\,|\,y, \mathbf{z}) \propto p(u)p(y\,|\,u, \mathbf{z}). \qquad (10.15)$$

The relation (10.15) takes exactly the same form as (10.11) with, of course, $p(u\,|\,\mathbf{t})$, the adjusted assessment for u, replaced by the direct assessment $p(u)$.

10.3 Calibrative distributions for the normal case

We confine ourselves throughout the remainder of this chapter to the case of linear regression with independently normally distributed errors. It is convenient to collect here the relevant distribution results preliminary to their application in the next section. The reader will find examples of other standard situations in the problems at the end of the chapter; see also Dunsmore (1968).

For the calibration experiment, whether natural or designed, we have the normal regression model for f_t, which sets

$$p(x\,|\,t, \boldsymbol{\theta}) = \text{No}(\alpha + \beta t, \tau),$$

where $\boldsymbol{\theta} = (\alpha, \beta, \tau)$; and for the future experiment f_u

$$p(y \mid u, \boldsymbol{\theta}) = \text{No}(\alpha + \beta u, \tau).$$

Since our interest is in the combination $\mu = \alpha + \beta u$ and τ we can, as in our discussion of example 9.2 in §9.2, summarise the data in the trial records $(t_1, x_1), \ldots, (t_n, x_n)$ constituting e by

$$m = \bar{x} + \hat{\beta}(u - \bar{t}), \quad v = \Sigma\{x_i - \bar{x} - \hat{\beta}(t_i - \bar{t})\}^2.$$

Then (m, v) satisfies the conditions of case 5 of table 2.3 with

$$\frac{1}{k} = \frac{1}{n} + \frac{(u - \bar{t})^2}{S(\mathbf{t}, \mathbf{t})}, \quad \nu = n - 2, \quad K = 1.$$

With a NoCh(b, c, g, h) prior density function on (μ, τ) we arrive at a Student form for the predictive distribution as defined in (10.13):

$$p(y \mid u, \mathbf{z}) = \text{St}\left\{G, B, \left(1 + \frac{1}{C}\right)\frac{H}{G}\right\};$$

the particular form of this when we adopt the vague prior distribution for (μ, τ) is:

$$p(y \mid u, \mathbf{z}) = \text{St}\left\{\nu, m, \left(1 + \frac{1}{k}\right)\frac{v}{\nu}\right\}. \tag{10.16}$$

For a natural calibration experiment the assumption of normality for $p(t \mid \psi)$, say No(λ, ρ), together with a NoCh prior on (λ, ρ) and the data $t_1, \ldots, t_n$ from e, leads again through a simple application of case 5 of table 2.3 to a Student form for $p(u \mid \mathbf{t})$ as defined in (10.12). The particular form of this when we adopt the vague prior distribution for (λ, ρ) is:

$$p(u \mid \mathbf{t}) = \text{St}\left(n - 1, \bar{t}, \left(1 + \frac{1}{n}\right)\frac{S(\mathbf{t}, \mathbf{t})}{n - 1}\right). \tag{10.17}$$

Substituting for $p(u \mid \mathbf{t})$ and $p(y \mid u, \mathbf{z})$ in (10.11) we see that $p(u \mid y, \mathbf{z})$ is proportional to the product of two Student-type density functions, and in general this does not reduce to any standard distribution. This, however, does not detract from the applicability of the method since it is an elementary computational exercise, trivial even on a small computer, to obtain the appropriate conversion factor to yield the appropriate density function $p(u \mid y, \mathbf{z})$. This technique is illustrated in the first of the applications of the next section.

For a designed calibration experiment the source of information on which to base the choice of $p(u)$ in (10.15) is outwith the calibration experiment. Its choice must in some sense reflect the opinion which governed the selection of the t-values $t_1, \ldots, t_n$. Again the product $p(u)p(y \mid u, \mathbf{z})$ is unlikely to yield a standard density function but our previous remarks regarding simple computation apply equally here.

Much research effort has gone into attempts to find some suitable $p(u)$ which will yield a $p(u|y, \mathbf{z})$ of standard form. While we feel that this search for tractability is not really necessary in view of the simplicity of the computational problem we report on it briefly in § 10.5. Since much of the argument will there centre on $p(y|u, \mathbf{z})$ as a function of u we set this out here in order to prepare the way for § 10.5. After some algebraic rearrangement we can express $p(y|u, \mathbf{z})$, in so far as it contains u, as being proportional to

$$\frac{\left\{n-3+\dfrac{n(n-3)}{(n+1)S(\mathbf{t},\mathbf{t})}(u-\bar{t})^2\right\}^{(n-2)/2}}{\left\{(n-2)+\dfrac{(u-u_I)^2}{Q(\mathbf{z},y)}\right\}^{(n-1)/2}}, \tag{10.18}$$

where

$$u_I = \bar{t}+\frac{S(\mathbf{t},\mathbf{x})}{S(\mathbf{x},\mathbf{x})}(y-\bar{x}), \tag{10.19}$$

and

$$Q(\mathbf{z},y) = \frac{vS(\mathbf{t},\mathbf{t})}{(n-2)S(\mathbf{x},\mathbf{x})}\left(1+\frac{1}{n}+\frac{(y-\bar{x})^2}{S(\mathbf{x},\mathbf{x})}\right). \tag{10.20}$$

All the results so far have been concerned with a single performance of the future experiment f_u. There are circumstances where it may be necessary to perform several replicates of f_u to obtain a precise enough calibrative distribution. We thus adjust our analysis to cover the case where K replicates of f_u have been performed with outcomes $y_1, \ldots, y_K$. By sufficiency arguments for $\mu = \alpha + \beta u$ and τ we can clearly restrict attention to a condensed future experiment which records (M, V), where

$$M = \bar{y}, \quad V = \Sigma(y_i-\bar{y})^2. \tag{10.21}$$

The distributions of M and V, for given u and $\boldsymbol{\theta} = (\alpha, \beta, \tau)$, are independent and take the following forms:

$$p(M|u,\boldsymbol{\theta}) = \text{No}(\mu, K\tau), \quad p(V|u,\boldsymbol{\theta}) = \text{Ch}(K-1,\tau). \tag{10.22}$$

We have thus again case 5 of table 2.3 and can immediately arrive at the predictive distribution for (M, V). The particular form corresponding to the vague prior on (μ, τ) is the following:

$$p(M, V|u, m, v) = \text{StSi}\left\{n-2; m, \left(\frac{1}{k}+\frac{1}{K}\right)\frac{v}{\nu}; K-1, v\right\}, \tag{10.23}$$

where

$$\frac{1}{k} = \frac{1}{n}+\frac{(u-\bar{t})^2}{S(\mathbf{t},\mathbf{t})}, \quad \nu = n-2, \tag{10.24}$$

to which we apply whatever prior distribution for u is appropriate. We shall see an application of this result in the analysis of example 10.2 in §10.4.

Again we record the form of $p(M, V|u, m, v)$ as a function of u, so that we can readily study the search for a tractable $p(u)$ in §10.5. In this case the function is proportional to

$$\frac{\left\{n+K-4+\dfrac{nK(n+K-4)}{(n+K)S(\mathrm{t},\mathrm{t})}(u-\bar{t})^2\right\}^{(n+K-3)/2}}{\left\{n+K-3+\dfrac{(u-u_I)^2}{Q(\mathrm{z},y)}\right\}^{(n+K-2)/2}}, \tag{10.25}$$

where in this case

$$u_I = \bar{t}+\frac{S(\mathrm{t},\mathrm{x})}{S(\mathrm{x},\mathrm{x})+V}(M-\bar{x}), \tag{10.26}$$

and

$$Q(\mathrm{z},y) = \frac{(v+V)S(\mathrm{t},\mathrm{t})}{(n+K-3)\{S(\mathrm{x},\mathrm{x})+V\}}\left\{\frac{1}{K}+\frac{1}{n}+\frac{(M-\bar{x})^2}{S(\mathrm{x},\mathrm{x})+V}\right\}. \tag{10.27}$$

10.4 Two applications of calibrative distributions

We now apply the results of the preceding two sections to the two motivating examples of §10.1.

Example 10.1 (continued)

Measuring water content of soil specimens. Straightforward regression calculations yield the following values in the notation of §10.3:

$$n = 16, \quad \hat{\alpha} = -1.60, \quad \hat{\beta} = 0.770,$$

$$\bar{t} = 26.3, \quad S(\mathrm{t},\mathrm{t}) = 1066.5, \quad v = 29.6.$$

We recall that this is a case of a natural calibration experiment so that we require the two factors of (10.11). These are given by substitution of the above values in (10.16) and (10.17):

$$p(y|u,\mathrm{z}) = \mathrm{St}\{14, -1.60+0.770u, 2.24+0.00198\,(u-26.3)^2\}, \tag{10.28}$$

$$p(u|\mathrm{t}) = \mathrm{St}(15, 26.3, 75.5). \tag{10.29}$$

The numerical method of constructing the calibrative density function for a given y is first to compute the product (10.11) for $u = b, b+h, b+2h, \ldots,$ $b+(N-1)h$, where $b+Nh = c$, and where (b, c) can easily be chosen to provide a sufficiently wide range for u, and h chosen sufficiently small to make

$$h \sum_{u=b}^{b+(N-1)h} p(u|\mathbf{t})p(y|u,\mathbf{z})$$

or some more sophisticated numerical integration formula, such as Simpson's rule, a good approximation to

$$\int_T p(u|\mathbf{t})p(y|u,\mathbf{z})du.$$

The calibrative density function can then be simply graphed for the specification

$$p(u|y,\mathbf{z}) = \frac{p(u|\mathbf{t})p(y|u,\mathbf{z})}{h\,\Sigma p(u|\mathbf{t})p(y|u,\mathbf{z})} \quad (u = b, b+h, \ldots, c). \quad (10.30)$$

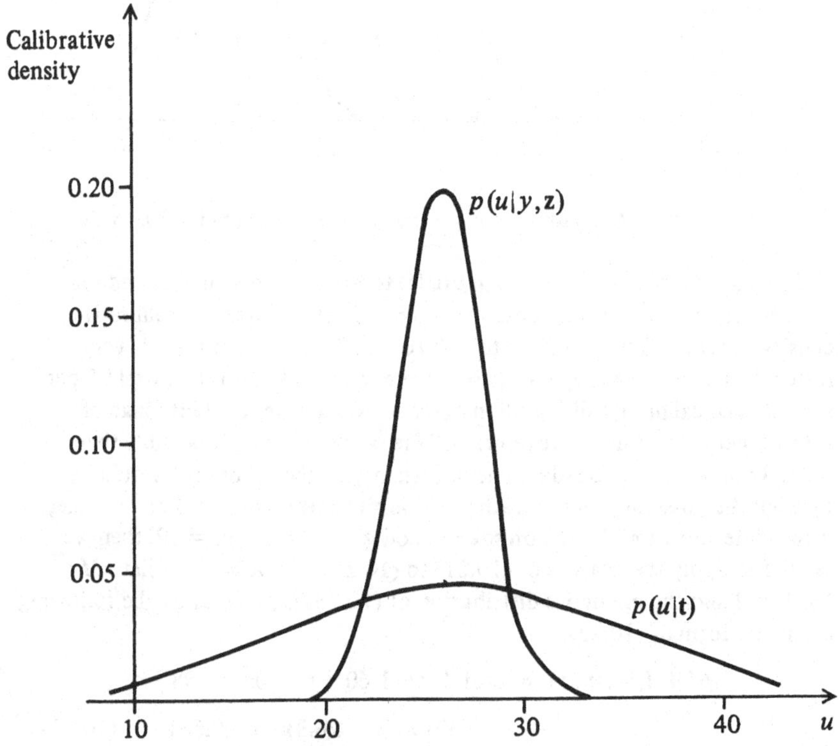

Fig. 10.3 Change from $p(u|\mathbf{t})$ to $p(u|y, \mathbf{z})$ for an observed on-site determination of water content $y = 18$.

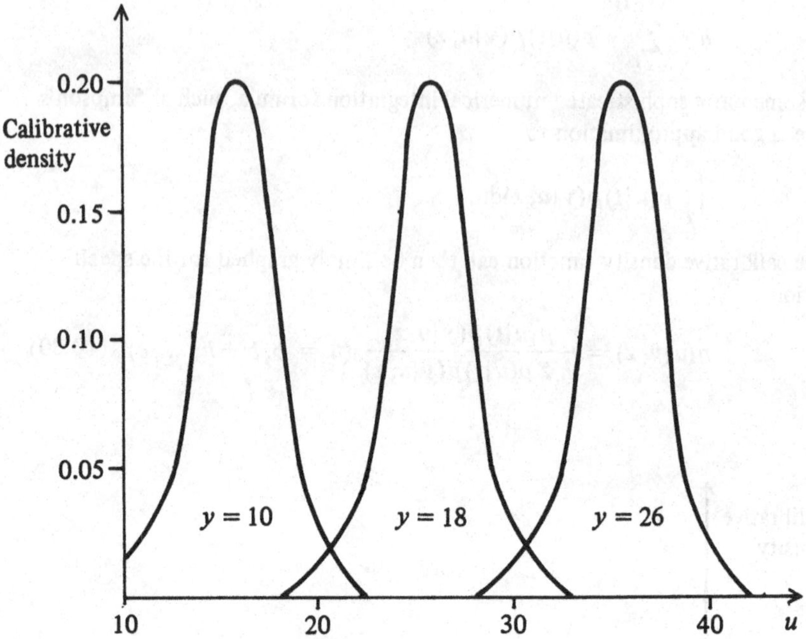

Fig. 10.4 Calibrative density functions corresponding to $y = 10, 18, 26$.

Fig. 10.3 shows the change from $p(u|\mathbf{t})$ to $p(u|y, \mathbf{z})$ for an observed on-site determination of water content $y = 18$. Fig. 10.4 shows the calibrative density functions corresponding to $y = 10, 18, 26$. Note that there is very little difference in the shape of these three curves, and that the central 95 per cent of each calibrative distribution extends over a water content range of about 8 per cent. If no improvement of the on-site method is possible then the only opportunity to reduce the uncertainty of the calibration would be through the possibility of replicating the on-site determination. For example, if two determinations by the on-site method give $y_1 = 17, y_2 = 19$, then we require to apply the analysis of (10.21) to (10.24) with $K = 2$. We have $M = 18$, $V = 2$ and the predictive distribution of (10.28) is replaced by the following particular form of (10.23):

$$p(M, V|u, m, v) = \text{StSi}\{14; -1.60 + 0.770u, 1.188 + 0.00198(u - 26.3)^2; 1, 29.56\}. \quad (10.31)$$

We can again apply the technique of (10.30) to obtain the calibrative distribution. Fig. 10.5 shows the extent to which the calibration is improved when two replicates are used instead of just one.

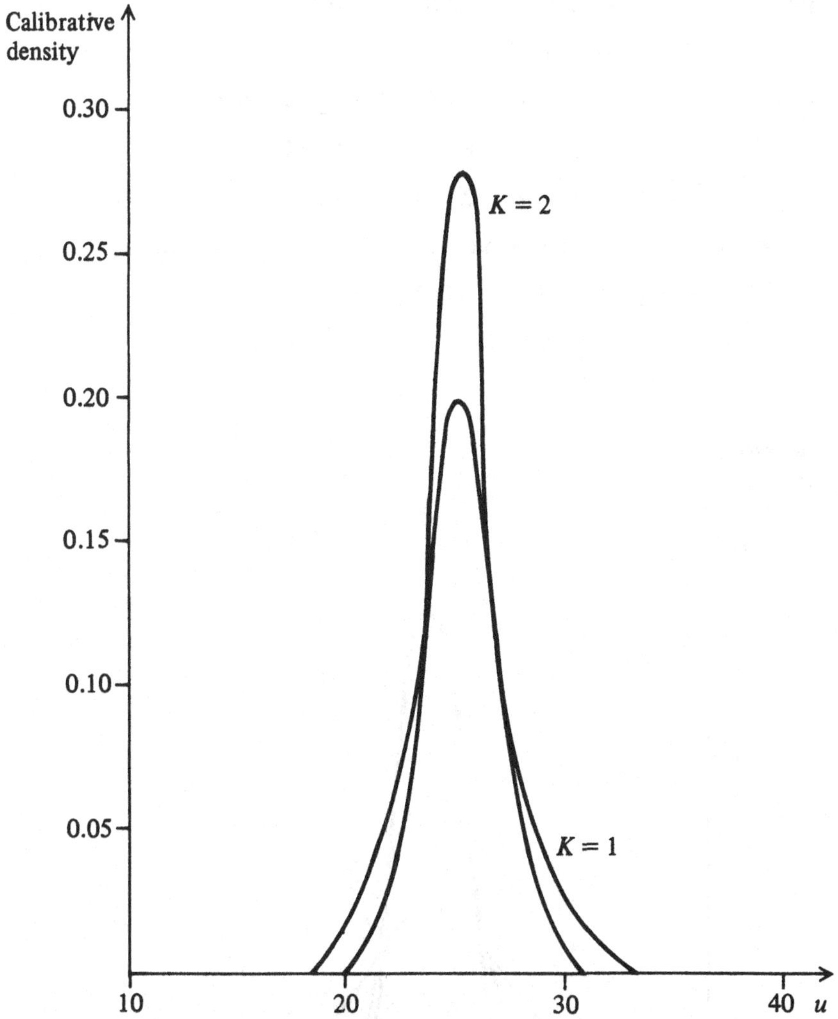

Fig. 10.5 Comparison of calibrative density functions for one and two replicates.

Example 10.2 (continued)

Calibration of an autoanalyser. As we have already pointed out in § 10.1 this is an example of a designed calibration experiment and so the calibrative density function is evaluated by use of (10.15). In fact we are given a firm basis for the specification of $p(u)$ in the information that the samples presented for analysis are likely to have enzyme concentrations which are normally distributed with mean 4.6 and standard deviation 0.8. Hence

$$p(u) = \text{No}(4.6, 1/0.64). \tag{10.32}$$

The calibration experiment provides the following regression values,

$$n = 27, \quad \hat{\alpha} = -0.0624, \quad \hat{\beta} = 0.814,$$

$$\bar{t} = 4.6, \quad S(\mathbf{t}, \mathbf{t}) = 28.80, \quad v = 0.223,$$

from which we construct the predictive density function from (10.16):

$$p(y\,|\,u, \mathbf{z}) = \text{St}\{25, -0.0624 + 0.814u,$$
$$0.00926 + 0.000310\,(u - 4.6)^2\}. \tag{10.33}$$

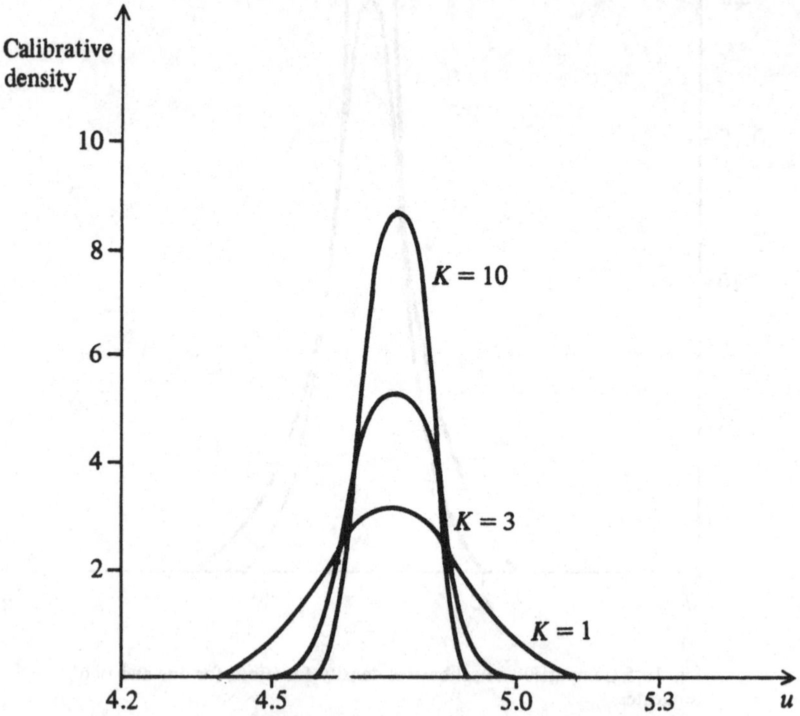

Fig. 10.6 Calibrative density functions for the autoanalyser problem.

We can then apply a computational technique similar to (10.30) to obtain the calibrative density function corresponding to $y = 3.8$. Fig. 10.6 shows this density function. The calibrative density function corresponding to other

y-values in the range likely to occur is very similar in shape. If this degree of reliability available from one aliquot is not satisfactory then we can investigate along the lines of (10.21)–(10.24) the consequences of using more aliquots. Thus we could readily see how many aliquots are required to provide a specified degree of reliability. To indicate the extent of such an increased reliability we show also in fig. 10.6 the calibration density functions corresponding to the following cases:

$$K = 3, \quad M = 3.8, \quad V = 0.02;$$

$$K = 10, \quad M = 3.8, \quad V = 0.09.$$

10.5 The search for tractable prior distributions

Suppose that in constructing the calibrative density function of (10.11) or of (10.15) we adopt an entirely mathematical approach and specify $p(u\,|\,\mathbf{t})$ or $p(u)$ in such a way that $p(u\,|\,y, \mathbf{z})$ turns out to be a standard function. While such an approach may be considered a rearrangement of priorities, a case of the end justifying the means, some practical justification for such a choice can be supplied in certain situations. For the normal case a choice of a tractable prior is influenced by the expression (10.18) for $p(y\,|\,u, \mathbf{z})$ as a function of u. The denominator of (10.18) depends on y, the outcome of the future experiment. Since our choice of $p(u\,|\,\mathbf{t})$ or $p(u)$ is made prior to the future experiment it seems unreasonable to allow $p(u\,|\,\mathbf{t})$ or $p(u)$ to depend on y. It is therefore impossible to eliminate the denominator from (10.18) and we turn our attention to the numerator. If we take $p(u\,|\,\mathbf{t})$ or $p(u)$ to be

$$\mathrm{St}\left\{n-3, \bar{t}, \left(1+\frac{1}{n}\right)\frac{S(\mathbf{t},\mathbf{t})}{n-3}\right\} \qquad (10.34)$$

then either (10.11) or (10.15) gives the very simple result that

$$p(u\,|\,y, \mathbf{z}) = \mathrm{St}\{n-2, u_I, Q(\mathbf{z}, y)\}, \qquad (10.35)$$

where u_I is as given by (10.19) and $Q(\mathbf{z}, y)$ by (10.20).

It now remains to examine whether there is any non-mathematical justification for the choice of (10.34). When dealing with a natural calibration experiment in § 10.3 we saw that $p(u\,|\,\mathbf{t})$ there takes the form (10.17). Comparison of (10.17) with (10.34) shows that the tractable prior distribution will be a reasonable approximation to (10.17) when n is reasonably large. For the water content calibration experiment fig. 10.3 provided the calibrative density function for $y = 18$. The tractable form (10.35) which here takes the particular form

$$\mathrm{St}(14, 25.5, 3.61)$$

is in fact indistinguishable in fig. 10.3 from this calibrative density function.

For a designed calibration experiment any justification is less direct. The mean and variance of the suggested prior distribution (10.34) are $\bar{t}$ and $(n+1)S(\mathrm{t},\mathrm{t})/\{n(n-5)\}$. If the designer of the calibration experiment has chosen the t-values $t_1, \ldots, t_n$ so that their mean and variance reflect the pattern of t-values he expects in the future then for reasonably large n the distribution (10.34) may give a satisfactory approximation. It should be emphasised, however, that there is no guarantee that such a design for the calibration experiment is optimum in the sense that it produces the most precise calibration distributions. For the enzyme calibration problem of § 10.4 the one-aliquot calibrative density function for $y = 3.8$ already shown in fig. 10.6 is again indistinguishable from the tractable calibrative density function (10.35) which here takes the particular form

$$\mathrm{St}(25, 4.74, 0.0138).$$

For the K-replicate future experiment discussed in § 10.3 it is clear from (10.25) that tractability arises from the specification of $p(u\,|\,\mathrm{t})$ or $p(u)$ as

$$\mathrm{St}\left\{n+K-4, \bar{t}, \left(\frac{1}{n}+\frac{1}{K}\right)\frac{S(\mathrm{t},\mathrm{t})}{n+K-4}\right\}. \tag{10.36}$$

Then, from either (10.11) or (10.15),

$$p(u\,|\,y, \mathrm{z}) = \mathrm{St}\{n+K-3, u_I, Q(\mathrm{z}, y)\}. \tag{10.37}$$

Unfortunately the choice of (10.36) as prior density function appears to be fairly arbitrary from a practical viewpoint. The prior mean and variance are given by $\bar{t}$ and $(n+K)S(\mathrm{t},\mathrm{t})/\{nK(n+K-6)\}$. Whilst in certain circumstances the mean of $\bar{t}$ may be satisfactory, the variance is harder to justify since it depends on K, the number of future replicates. Indeed we have the rather paradoxical situation that the more future observations there are (that is, the larger K is), the smaller is the prior variance. We see therefore that in this search for tractability there are certain difficulties involved in the case where $K > 1$. We reiterate our comment in § 10.3 that tractability in these calibration problems can lead to an unnecessary departure from reality.

10.6 Calibration under a utility structure

A statistical decision theory model for calibration has the following components.

Parameter space. The set of unknown parameters is the index set T for the class F of possible experiments, the unknown index u playing the role of the unknown state of nature.

Action set. The action set A is the set T for point calibration or a class of subsets of T for interval calibration.

Utility function. It seems sensible to define a utility function $U(a, u)$ which attaches a utility to each combination of possible calibration point or set a and each possible index u. The basic problem is to maximise the expected utility $U(y, a)$ of having obtained results $\mathbf{z}$ from the informative experiment and y from the 'future' experiment, and of having selected a region $a \subset T$.

We have that

$$U(y, a) = \int_T U(a, u) p(u \mid y, \mathbf{z}) du, \tag{10.38}$$

where $p(u \mid y, \mathbf{z})$ is the calibrative density function. Since we have full control over the selection of a we choose the index or interval which maximises $U(y, a)$, perhaps subject to some constraint. We note here a special feature of the analysis, namely the dependence of the expected utility on y. We are indeed using y to obtain information about θ additional to that already obtained from e. For some simple forms of $U(a, u)$, completely analogous to the utility functions considered in §3.1 and 3.3, we simply list in table 10.3 the optimum calibrations in terms of $p(u \mid y, \mathbf{z})$.

10.7 Some comparisons of methods

The calibrative distribution $p(u \mid y, \mathbf{z})$ is the cornerstone of the Bayesian model for the inverse prediction problem. It supplies the complete summary of our views on the value of u. From it we can provide either a point estimate (for example, the mean or mode of $p(u \mid y, \mathbf{z})$) or a set prediction (for example, the most plausible Bayesian interval of cover κ). Alternatively the Bayesian decision theory approach of §10.6 can provide such predictions.

The classical approach to the calibration problem has been subjected to much investigation (see references at the end of the chapter). We consider first the single-replicate future experiment situation for the normal case. The *classical* point prediction $\hat{u}$ is obtained simply by inverting the least squares regression line

$$y = \hat{\alpha} + \hat{\beta} u$$

to give

$$\begin{aligned} \hat{u} &= \bar{t} + \frac{(y - \bar{x})}{\hat{\beta}} \\ &= \bar{t} + \frac{S(\mathbf{t}, \mathbf{t})}{S(\mathbf{t}, \mathbf{x})} (y - \bar{x}). \end{aligned} \tag{10.39}$$

Table 10.3 *Optimal calibrations associated with some utility functions*

Utility function $U(a, u)$	Optimal prediction in terms of $p(u \mid y, z)$
Point prediction	
$\begin{cases} 1 & a-\epsilon \leqslant u \leqslant a+\epsilon \\ 0 & \text{otherwise} \end{cases}$	mode
$\begin{cases} -\xi(a-u) & u<a \\ -\eta(u-a) & u \geqslant a \end{cases}$	$\dfrac{\eta}{\xi+\eta}$ th quantile
$-(a-u)^2$	mean
Set prediction	
$\begin{cases} 1-\gamma(a_2-a_1) & a_1 \leqslant u \leqslant a_2 \\ -\gamma(a_2-a_1) & \text{otherwise} \end{cases}$	$\{u : p(u \mid y, z) > \gamma\}$
$\begin{cases} -\xi(a_1-u)-(a_2-a_1) & u<a_1 \\ -(a_2-a_1) & a_1 \leqslant u \leqslant a_2 \\ -\eta(u-a_2)-(a_2-a_1) & u>a_1 \end{cases}$	$\dfrac{1}{\xi}+\dfrac{1}{\eta} \leqslant 1 : \begin{cases} a_1{}^* = \dfrac{1}{\xi} \text{ th quantile} \\ a_2{}^* = \left(1-\dfrac{1}{\eta}\right) \text{ th quantile} \end{cases}$ $\dfrac{1}{\xi}+\dfrac{1}{\eta} > 1 : a_1{}^* = a_2{}^* = \dfrac{\eta}{\xi+\eta}$ th quantile

If one considers u as an unknown parameter then $\hat{u}$ is simply the maximum likelihood estimate.

The *inverse* estimator (10.19) given by

$$u_I = \bar{t} + \frac{S(\mathbf{t}, \mathbf{x})}{S(\mathbf{x}, \mathbf{x})}(y - \bar{x}) \tag{10.40}$$

has also been proposed. Formally this can be obtained by fitting a line $t = \gamma + \delta x$ to the data by means of least squares. This estimate has the advantage over $\hat{u}$ of having a finite mean square error, at least for $n \geqslant 4$.

Recall that u_I is the mean value of the calibrative distribution (10.37). From a Bayesian viewpoint the inverse estimator can be thought of as a shift of $\hat{u}$ towards the prior mean which gives the smallest adjustments to $\hat{u}$ when the data are most informative. We note that in all situations $|u_I - \bar{t}| \leqslant |\hat{u} - \bar{t}|$, with equality only in the case of an exact relationship. Thus the more informative the data the more we move from the prior mean towards the estimate $\hat{u}$.

Several authors have decried the use of u_I (for example, Berkson (1969), Williams (1969*a*), Halperin (1970). Martinelle (1970)), stating categorically that $\hat{u}$ is to be preferred to u_I despite the fact that $\hat{u}$ has an undefined mean and infinite mean square error.

Difficulties can also be encountered in the classical confidence interval approach although in most practical situations the analysis is straightforward. The classical confidence region is obtained from the fact that

$$\frac{y-\bar{x}-\hat{\beta}(u-\bar{t})}{\left\{1+\frac{1}{n}+\frac{(u-\bar{t})^2}{S(\mathrm{t},\mathrm{t})}\right\}^{1/2}} \bigg/ \left(\frac{v}{n-2}\right)^{1/2} \quad \text{is St}(n-2,0,1),$$

so that a 100 $(1-\alpha)$ per cent region is given by

$$\left\{u: \frac{(n-2)\hat{\beta}^2(\hat{u}-u)^2}{v\left\{1+\frac{1}{n}+\frac{(u-\bar{t})^2}{S(\mathrm{t},\mathrm{t})}\right\}} < \mathrm{F}(1,n-2;1-\alpha)\right\}.$$

This yields a confidence region of one of three possible forms which depend on the value of

$$R = \frac{(n-2)\hat{\beta}^2 S(\mathrm{t},\mathrm{t})}{v}. \tag{10.41}$$

We have the regions

$$\left.\begin{array}{ll}
\{u: u_1 < u < u_2\} & \text{if } R > \mathrm{F}(1,n-2;1-\alpha), \\
\{u: u \leqslant u_1 \text{ or } u \geqslant u_2\} & \text{if } \dfrac{\left(1+\frac{1}{n}\right)}{\left\{1+\frac{1}{n}+\frac{(\hat{t}-\bar{t})^2}{S(\mathrm{t},\mathrm{t})}\right\}} \\
& \quad \times \mathrm{F}(1,n-2;1-\alpha) < R \\
& \quad < \mathrm{F}(1,n-2;1-\alpha), \\
(-\infty,\infty) & \text{if } R < \dfrac{\left(1+\frac{1}{n}\right)}{\left\{1+\frac{1}{n}+\frac{(\hat{t}-\bar{t})^2}{S(\mathrm{t},\mathrm{t})}\right\}} \\
& \quad \times \mathrm{F}(1,n-2;1-\alpha),
\end{array}\right\} \tag{10.42}$$

where u_1, u_2 $(u_1 < u_2)$ are given by

$$\bar{t}+\frac{\hat{\beta}(y-\bar{x})}{w} \pm \frac{1}{w}\left[\frac{v}{n-2}\mathrm{F}(1,n-2;1-\alpha) \times \left\{\left(1+\frac{1}{n}\right)w+\frac{(y-\bar{x})^2}{S(\mathrm{t},\mathrm{t})}\right\}\right]^{1/2} \tag{10.43}$$

with

$$w = \hat{\beta}^2 - \frac{v\mathrm{F}(1, n-2; 1-\alpha)}{(n-2)S(\mathrm{t}, \mathrm{t})}.$$

There is thus a chance that the 100 (1 − α) per cent confidence region will be given by (−∞, ∞) and so be rendered useless. This unlimited region occurs when the hypothesis $\beta = 0$ cannot be rejected on the information available. Of course the probability of this may be very small, thus negating the importance of this contingency. In much the same way the fact that $\hat{u}$ has undefined mean and infinite mean square error may not be disastrous – the distribution of $\hat{u}$ may still be quite well behaved.

Hoadley (1970) suggests a third reason why $\hat{u}$ may be considered unsatisfactory. The data from *e* contain some information about the precision of $\hat{u}$. Thus if $\hat{u}$ is known to be unreliable, less weight or importance should be attached to it. This suggests that the Bayesian approach is more realistic.

A similar situation exists when we consider the case of the *K*-replicate future experiment discussed in §10.3. Here three point estimates have been proposed, namely

(i) the *classical* estimate

$$\hat{u} = \bar{t} + \frac{M - \bar{x}}{\hat{\beta}}; \tag{10.44}$$

(ii) the *inverse* estimate suggested by Krutchkoff (1967)

$$\begin{aligned} \tilde{u} &= \bar{t} + \frac{S(\mathrm{t}, \mathrm{x})}{S(\mathrm{x}, \mathrm{x})}(M - \bar{x}) \\ &= \bar{t} + \frac{S(\mathrm{t}, \mathrm{x})}{\hat{\beta}^2 S(\mathrm{t}, \mathrm{t}) + v}(M - \bar{x}); \end{aligned} \tag{10.45}$$

(iii) Halperin's (1970) family of *modified inverse* estimates

$$u(r) = \bar{t} + \frac{rS(\mathrm{t}, \mathrm{x})}{r\hat{\beta}^2 S(\mathrm{t}, \mathrm{t}) + v}(M - \bar{x}). \tag{10.46}$$

These compare with the mean u_I of the calibrative distribution (10.37) where

$$u_I = \bar{t} + \frac{S(\mathrm{t}, \mathrm{x})}{S(\mathrm{x}, \mathrm{x}) + V}(M - \bar{x}). \tag{10.47}$$

The balance of opinion favours (10.44) from the three classical estimates (10.44), (10.45) and (10.46). The point estimate u_I has one appealing property. The variation in $y_1, y_2, \ldots, y_K$, as measured by V, is incorporated in the estimate in such a way that the more variation there is in the future observations the nearer is the predictive mean to $\bar{t}$. From a Bayesian viewpoint with prior mean $\bar{t}$ this is an intuitively pleasing property for one should be less willing to change one's prior view if the additional data are very variable. None of the other estimates take this variation into account in obtaining a point estimate. Notice that

$$|u_I - \bar{t}| \leqslant |\tilde{u} - \bar{t}| \leqslant |\hat{u} - \bar{t}|,$$

so that u_I places less weight on $\hat{u}$ than $\tilde{u}$ does and is less removed from the sample mean $\bar{t}$ than $\tilde{u}$. If $r \geqslant 1$ (and one of the important cases considered by Halperin is where $r = K$) then we have further that

$$|u_I - \bar{t}| \leqslant |\tilde{u} - \bar{t}| \leqslant |u(r) - \bar{t}| \leqslant |\hat{u} - \bar{t}|.$$

As in the single replicate case difficulties can be encountered in obtaining a confidence interval. The extension from (10.42) is straightforward. Suffice it to say that for suitable cases, that is if

$$\frac{(n+K-3)\,\hat{\beta}^2 S(\mathrm{t},\mathrm{t})}{v+V} > \mathrm{F}(1, n+K-3; 1-\alpha),$$

the $100(1-\alpha)$ per cent confidence region is given by

$$\bar{t} + \frac{\hat{\beta}(M-\bar{x})}{W} \pm \frac{1}{W}\left[\frac{(v+V)}{(n+K-3)}\,\mathrm{F}(1, n+K-3; 1-\alpha) \times \left\{\left(\frac{1}{n}+\frac{1}{K}\right)W + \frac{(M-\bar{x})^2}{S(\mathrm{t},\mathrm{t})}\right\}\right]^{1/2} \qquad (10.48)$$

where

$$W = \hat{\beta}^2 - \frac{(v+V)\,\mathrm{F}(1, N+K-3; 1-\alpha)}{(n+K-3)S(\mathrm{t},\mathrm{t})}.$$

This interval is not positioned symmetrically about the classical estimate $\hat{u}$. The variation in the values $y_1, y_2, \ldots, y_K$ is now taken into account, and it can be seen that, for a given K, the more the variation the wider is the corresponding confidence interval.

Halperin's estimate $u(r)$ does not allow an exact interval estimate of u. He feels that in most practical situations there would be little to choose between $u(r)$ and $\hat{u}$, and so prefers $\hat{u}$ since an exact interval estimate is usually obtainable.

As a simple illustration to compare the methods we consider again example 10.1. The results of the informative experiment are shown in fig. 10.1. We will consider several different situations for the 'future' experiment in each of which $M = 18$, but where K and the variation in $y_1, y_2, \ldots, y_K$ alter. These specifications are listed in table 10.4.

Point prediction. The classical estimate $\hat{u}$ (10.44), the inverse estimate $\tilde{u}$ (10.45) and Halperin's estimate $u(r)$ (10.46) are invariant for all the

Table 10.4 *Several future experiments in example 10.1 with M = 18.*

Specification	1	2(i)	2(ii)	2(iii)	3(i)	3(ii)	3(iii)	3(iv)
K	1	2	2	2	3	3	3	3
y	18	17.5 19.5	17.0 19.0	16.0 20.0	17.8 18.0 18.2	17.6 17.8 18.6	17.0 18.0 19.0	16.0 19.0 19.0
V	0	0.50	2.0	8.0	0.08	0.56	2.00	6.00

Table 10.5 *Bayesian and classical 95 per cent set prediction*

Specification	Bayesian (10.49)	Classical (10.48)
1	(21.40, 29.56)	(21.22, 29.63)
2(i)	(22.61, 28.35)	(22.47, 28.39)
2(ii)	(22.54, 28.42)	(22.39, 28.46)
2(iii)	(22.30, 28.68)	(22.11, 28.74)
3(i)	(23.17, 27.78)	(23.05, 27.80)
3(ii)	(23.16, 27.80)	(23.04, 27.82)
3(iii)	(23.11, 27.86)	(22.98, 27.88)
3(iv)	(22.97, 28.00)	(22.82, 28.03)

specifications. We find that

$$\hat{u} = 25.44, \quad \tilde{u} = 25.48$$

and

r	1	2	5	10	50	100
$u(r)$	25.48	25.46	25.45	25.44	25.44	25.44

The only estimate which varies with K and V is u_I (10.47), and we obtain

Specification	1	2(i)	2(ii)	2(iii)	3(i)	3(ii)	3(iii)	3(iv)
u_I	25.48	25.48	25.48	25.49	25.48	25.48	25.48	25.49

We see that for given K the more variation there is in $y_1, y_2, \ldots, y_K$, the less we move away from the prior mean $\bar{t} = 26.30$.

Set Prediction. No difficulty arises in the classical confidence interval approach in this situation. In table 10.5 we give the 95 per cent intervals for each of the specifications. We also give the 95 per cent (shortest) Bayesian intervals, obtained from (10.37) and given by

$$u_I \pm \sqrt{[Q(z, y)]}\, t\,(n + K - 3; 1 - \tfrac{1}{2}\alpha). \qquad (10.49)$$

In each case we see that the Bayesian interval is the shorter.

Table 10.6 *Bayesian set predictions from utility functions*

γ	Specification 1	2(ii)	3(iii)	Bayesian cover	ξ
0.01	(19.83, 31.13)	(21.41, 29.56)	(22.19, 28.77)	0.99	200
0.05	(21.48, 29.48)	(22.59, 28.38)	(23.13, 27.83)	0.95	40
0.10	(22.30, 28.68)	(23.18, 27.79)	(23.61, 27.35)	0.88	17
0.20	(23.30, 27.66)	(23.90, 27.06)	(24.19, 26.77)	0.73	7
0.30	(24.12, 26.83)	(24.49, 26.47)	(24.68, 26.29)	0.51	4

If the Bayesian decision theory approach is used the specification of γ or ξ, η in the utility functions in table 10.3 is of importance. Table 10.6 shows the intervals obtained for three of the specifications from table 10.4 for a selection of suitable values of γ. In column 5 the (approximate) predictive probabilities of the selected regions are shown. (These can easily be found by use of (10.37) and table 9 in Pearson and Hartley, 1966.) It so happens that, for a given γ, the predictive probabilities are almost identical for each specification. This is because there is not very much difference between the t-distributions with 14, 15 and 16 degrees of freedom.

Suppose we use the last utility of table 10.3. For simplicity we assume $\xi = \eta$. The approximate values of ξ which correspond to the intervals obtained in table 10.6 are shown in column 6 of that table.

Basically the imposition of the extra structure afforded by the utility functions leads to different ways of summarising the predictive density function $p(u|y, z)$.

10.8 An application to antibiotic assay

The analysis of example 1.6 (calibration by biological assay) may now be completed with the already familiar tools developed earlier in this chapter. We need draw attention only to some particular aspects.

First from the scatter diagram of fig. 1.2 we see that the regression of clearance diameter on the logarithm of concentration is not linear over the whole range of concentrations explored. Some alternatives are open to us. For example, we may attempt to adopt a suitable non-linear form for the regression, or try to obtain some transformation of the clearance diameters which will linearise the regression, or restrict our operations to the existing linear stretch of fig. 1.2. It is often convenient in assay problems to adopt the third alternative since there is usually the opportunity of investigating successive dilutions of a specimen and so obtaining a particular dilution lying within the linear part of the concentration range. For our particular problem we confine attention to the range $t = 1$ to $t = 3$ for $\log_2$ (concentration). The informative experiment e then consists of 60 trial records (the central three columns of table 1.5), and the usual regression calculations in the notation of §10.3 give:

$$n = 60, \quad \hat{\alpha} = 15.1, \quad \hat{\beta} = 1.94,$$

$$\bar{t} = 2.00, \quad S(t, t) = 40.0, \quad v = 48.3.$$

The calibration experiment e is here designed and so we have to choose a sensible $p(u)$ from other considerations. If preliminary dilutions have convinced us that the current specimen has a concentration within the range $1 \leqslant u \leqslant 3$ then we should confine $p(u)$ to this range; the uniform distribution over this range may then be a reasonable basis for calibration. If, however, there is the possibility that the concentration of the specimen lies outside the range $1 \leqslant u \leqslant 3$ then $p(u)$ must reflect this possibility. There is indeed a certain attraction in this second alternative since it provides a useful form of monitoring whether further dilution is required. We therefore adopt this alternative and assume that $p(u)$ is the improper uniform prior over the real line. We can thus set $p(u) = 1$ in (10.15). If we construct the calibrative density function $p(u \mid y, z)$ on this basis and find that it does not assign high plausibility to the calibration range $1 \leqslant u \leqslant 3$ then we require further dilution or possibly have over-diluted, and the calibrative density function will indicate which of these is the appropriate conclusion. In other words, we can construct a rule for reassaying the specimen at a different dilution of the following form: if

$$\int_1^3 p(u \mid y, z)\, du > c$$

use the calibrative density function $p(u \mid y, z)$ for the assay; otherwise reassay at a different dilution.

We now proceed to answer the three questions posed in example 1.6 on the basis of the vague prior on (α, β, τ) or equivalently on (μ, τ) where $\mu = \alpha + \beta u$.

(1) In the notation and terminology of § 10.3 we have a single replicate of f_u giving an observation $y = 19$. The predictive density function required for (10.15) is then of the form (10.16) and using the regression calculations already made we obtain

$$p(y \mid u, z) = \mathrm{St}\{58, 15.1 + 1.94u, 0.847 + 0.0208\,(u - 2.00)^2\}.$$

Application of the technique used in (10.30) with $y = 19$ gives the calibrative density function shown in fig. 10.7.

(2) For this calibration we have 3 clearance diameters so that in the notation and terminology of (10.21)–(10.24) we have

$$K = 3, \quad M = 19.0, \quad V = 1.50,$$

and

$$p(M, V \mid u, m, v) = \mathrm{StSi}\{58; 15.1 + 1.94u, 0.292 + 0.0208(u - 2)^2; 2, 48.3\}.$$

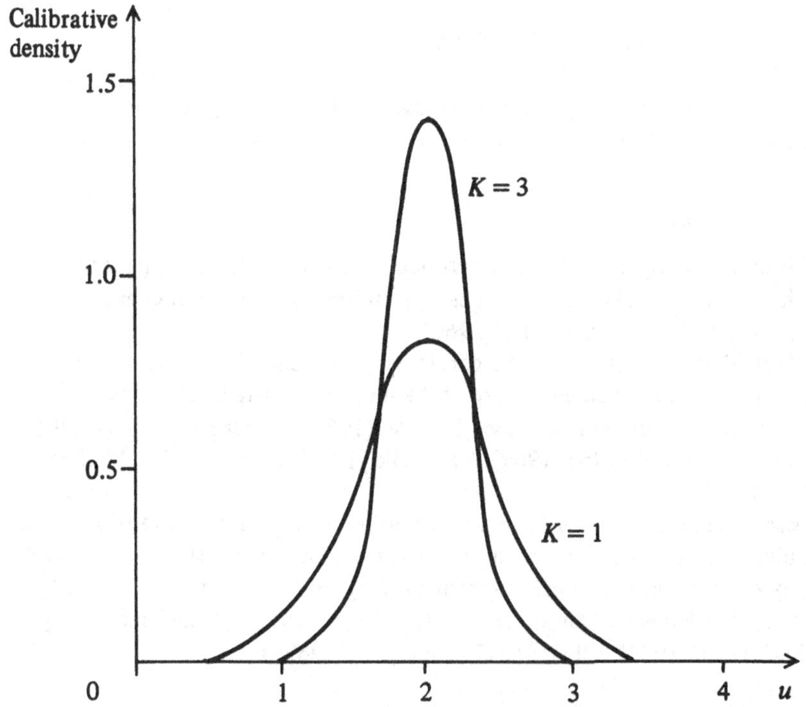

Fig. 10.7 Calibrative density functions for single clearance diameter 19, and for three clearance diameters 18.0, 19.5, 19.5.

Again it is easy to apply (10.30) for the given (M, V) to obtain the more precise calibrative density function also shown in fig. 10.7.

(3) To resolve this problem we have to determine how large K must be to ensure that the resulting calibrative density is highly concentrated round its mode. If we interpret 'reasonably' to be 90 per cent then we would require 90 per cent of the plausibility assigned by the calibrative density function to be within 10 per cent of its modal concentration, or in terms of the $\log_2$ (concentration) scale within $\log_2 0.9 = -0.15$ below and within $\log_2 1.1 = 0.14$ above the modal u. To investigate this we can easily apply the graphing technique to representative data based on different K, say with

$$M = 19, \quad V = (K-1)\frac{v}{n-2}$$

until we obtain the desired accuracy, in the present case until we obtain by numerical integration

$$\int_{1.85}^{2.14} p(u\,|\,y, z)\,du > 0.90.$$

By this technique we find that it is necessary to take K as big as 60 to obtain such accuracy and this is unlikely to be a practical proposition.

History

Calibration is a topic which many text books containing details of regression methods tend to avoid. Two which escape this criticism are Bowker and Lieberman (1959) and Brownlee (1960).

Krutchkoff (1967) provoked a considerable amount of controversy in his paper in which he proposed the use of the inverse estimate on the evidence provided by a simulation study; see Berkson (1969), Halperin (1970), Hoadley (1970), Krutchkoff (1968, 1969), Martinelle (1970), Shukla (1972), Williams (1969*a*).

Eisenhart (1939) had earlier discussed both the classical and inverse estimates for bivariate situations and concluded that it was clear which should be chosen in any circumstance, the decision resting simply on the nature of $t_1, t_2, \ldots, t_n$.

Further references: Dunsmore (1968), Kalotay (1971), Mandel and Linnig (1957), Oden (1973), Scheffé (1973), Williams (1969*b*).

Problems

10.1 Complete the analysis of Problem 1.5.

10.2 In a new production process items arrive at an inspection point according to a Poisson process with parameter ψ. The inspector either passes the item or rejects the item, in which case he channels it back for reprocessing. Suppose the individual items each have probability θ of being accepted. Little is known initially about θ and ψ. A count is made both of acceptable items and scrapped items during n trial eight-hour shifts. Suppose that in future only the acceptable items produced during a shift are counted. Can you estimate the number of scrapped items?

10.3 In a medical experiment testing the reactions of patients the number x of current responses given in a fixed period of time is related to the amount t of a stimulus in the patient's bloodstream. Suppose that $p(x\,|\,t, \theta)$ is $\text{Po}(t\theta)$. It is difficult to evaluate t and in future the amount of stimulus is to be estimated from the number of correct responses which the patient gives. To obtain information the experiment is simulated by injecting a series of patients with the dosages shown below and recording the numbers of correct responses.

For a new patient who records 5 correct answers what can be said about the level of stimulus in his bloodstream if initially it was thought equally likely to be any value in the range 0 to 1 ml?

t (ml) :	0.1	0.2	0.3	0.4	0.5	0.6	0.7	0.8	0.9
x :	2	3	2	4	4	6	6	8	7

10.4 An inexpensive quick chromatographic method for determining the excretion rate (mg/24h) of a certain steroid metabolite in urine has been developed. It is hoped that this method may in future replace the long and costly, though accurate, bioassay technique currently used. The considerable past experience of bioassays has shown that the excretion rates analysed are approximately normally distributed with mean 2 mg/24h and standard deviation 0.5 mg/24h. To explore the possibilities of the new method, aliquots from a number of urine samples are available. The experimenter has made bioassay determinations on one aliquot from each urine sample and selected a subset which he felt gave adequate coverage of the range of excretion rates. Three other aliquots from each urine sample of this subset were then assigned to the chromatographic method and the results are shown in the table.

Explore the possibilities of using the chromatographic method in future.

Serial no. i of urine sample	Excretion rate (mg/24h) Bioassay method	Chromatographic method
1	0.50	0.80, 0.88, 0.98
2	1.00	1.07, 1.10, 1.10
3	1.20	1.20, 1.23, 1.35
4	1.40	1.36, 1.48, 1.49
5	1.60	1.52, 1.53, 1.56
6	1.80	1.63, 1.72, 1.82
7	2.00	1.76, 1.80, 1.88
8	2.20	1.95, 2.00, 2.02
9	2.40	2.01, 2.04, 2.18
10	2.60	2.16, 2.28, 2.29
11	2.80	2.31, 2.40, 2.42
12	3.30	2.45, 2.51, 2.52
13	3.50	2.82, 2.94, 3.01

10.5 The gain in weight of a rat depends on the amount of a certain vitamin in its diet. A random sample of 25 batches of feed are known to contain concentrations $t_1, t_2, \dots, t_{25}$ of the vitamin. In the informative experiment 25 rats were used, each rat being assigned to a different batch, and the weight gains $x_1, x_2, \dots, x_{25}$ shown below are noted after each has received the same amount of feed. The concentration of the vitamin in a further batch is unknown. The weight gains of K rats fed from this batch with the same amount of feed as before are shown below for several specifications, each of which has $M = 8.4$. Compare the different calibrations of § 10.7 for the specifications given.

Amount of vitamin (mg)	21.5	18.7	23.5	16.1	24.8	16.5	21.1
Gain in weight (g)	8.4	7.3	8.9	7.2	9.5	6.4	8.0
Amount of vitamin	23.8	18.9	14.1	24.9	18.2	16.5	23.4
Gain in weight	8.4	6.9	5.2	8.9	6.2	7.0	8.6
Amount of vitamin	21.7	20.7	19.9	24.0	14.6	20.9	20.1
Gain in weight	8.1	8.0	7.8	8.9	8.2	5.1	7.7
Amount of vitamin	22.5	22.9	20.4	15.4			
Gain in weight	9.0	8.0	7.2	6.4			

Specification	1	2(i)	2(ii)	2(iii)	3(i)	3(ii)	3(iii)	3(iv)
K	1	2	2	2	3	3	3	3
$y = (y_1, \dots, y_K)$	8.4	8.2	7.9	7.4	8.3	8.0	7.7	7.6
		8.6	8.9	8.4	8.4	8.2	8.0	7.6
					8.5	9.0	9.5	10.0

10.6 In the bioassay of an antibiotic by the clearance circle technique the following table gives the clearance diameters associated with standard antibiotic preparations of known dilution, together with the six clearance diameters from each of two antibiotic specimens of unknown 'dilutions'. What can usefully be inferred about these unknown dilutions?

Standard antibiotic preparations

Dilution	Clearance diameters (mm)
1	24, 22, 26, 25
$\frac{1}{2}$	17, 16, 14, 18
$\frac{1}{4}$	10, 14, 12, 14
$\frac{1}{8}$	8, 8, 10, 7
$\frac{1}{16}$	6, 5, 7, 6
$\frac{1}{32}$	4, 4, 5, 5

Antibiotic specimens

Specimen no	Clearance diameters
1	15, 13, 12, 14, 15, 15
2	10, 10, 11, 8, 8, 9

10.7 The calibration problem associated with (10.21)–(10.24) envisages an informative experiment e with m and v distributed independently as $\mathrm{No}(\mu, k\tau)$ and $\mathrm{Ch}(\nu, \tau)$ and a 'future' experiment f_u with M and V distributed independently as $\mathrm{No}(\mu, K\tau)$ and $\mathrm{Ch}(K-1, \tau)$. Show that the calibrative density function is exactly the same as that obtained from an informative experiment e' with m and $v + V$ distributed independently as $\text{No}(\mu, k\tau)$ and $\text{Ch}(\nu + K - 1, \tau)$ and a future experiment f_u' yielding only M which is $\text{No}(\mu, K\tau)$. Can you provide an intuitive argument for this result?

11
Diagnosis

11.1 The nature of a diagnostic problem

Although for this chapter we have used a title which has often a medical connotation the problem arises in many other fields – for example, in the diagnosis of a fault in a complex industrial process, in categorising an archaeological or anthropological specimen. From an expository point of view, however, the nature of a diagnostic problem is most easily described, and the corresponding theory is best developed, within the context of a specific situation. For this purpose we have selected a medical problem concerning the differential diagnosis of three forms or types of a particular syndrome on the basis of two diagnostic tests or observable features. We have deliberately selected this three-type two-feature problem because it allows the maximum exploitation of diagrammatic means of expressing concepts and analyses. All the concepts and analyses carry over straightforwardly into higher dimensional problems. Indeed the introductory illustrative problem which we now present is a subproblem extracted from a larger real one.

Example 11.1

Differential diagnosis of Cushing's syndrome. Cushing's syndrome is a rare hypersensitive disorder associated with the over-secretion of cortisol by the adrenal cortex. For illustrative purposes we confine ourselves here to three 'types' of the syndrome, those types in which the cause of this over-secretion is actually within the adrenal gland itself. The types are

a: adenoma,
b: bilateral hyperplasia,
c: carcinoma,

and we investigate the possibilities of distinguishing the types on the basis of two observable 'features', the determination by paperchromatography of the urinary excretion rates (mg/24h) of two steroid metabolites, tetrahydrocortisone and pregnanetriol. Table 11.1 gives these rates for 21 patients with Cushing's syndrome, who in the past have all been operated on and the particular type a, b or c histopathologically determined. For each of these past cases the verified type of the syndrome and the feature vectors can be represented by a

Table 11.1 *Urinary excretion rates (mg/24h) of two steroid metabolites for 21 patients with Cushing's syndrome*

Case no.	Tetrahydrocortisone	Pregnanetriol
*a*1	3.1	11.70
*a*2	3.0	1.30
*a*3	1.9	0.10
*a*4	3.8	0.04
*a*5	4.1	1.10
*a*6	1.9	0.40
*b*1	8.3	1.00
*b*2	3.8	0.20
*b*3	3.9	0.60
*b*4	7.8	1.20
*b*5	9.1	0.60
*b*6	15.4	3.60
*b*7	7.7	1.60
*b*8	6.5	0.40
*b*9	5.7	0.40
*b*10	13.6	1.60
*c*1	10.2	6.40
*c*2	9.2	7.90
*c*3	9.6	3.10
*c*4	53.8	2.50
*c*5	15.8	7.60

point in a two-dimensional diagram, as in fig. 11.1. In this diagram logarithmic scales have been used to accommodate the large proportional differences that occur in these excretion rates. This diagram can be thought of as representing past experience or past case records; the results indeed constitute our informative experiment *e*. In fig. 11.1 there is clearly some degree of separation of the three types and so some hope that the excretion rates will be of some diagnostic value for future cases.

If we denote by f_t the 'experiment' which records the two urinary excretion rates of a patient who has type *t*, then when we know that we are dealing with a Cushing patient we are performing an experiment from the class

$$F = \{f_t : t \in T\}, \text{ where } T = \{a, b, c\},$$

of possible experiments. We thus can write

$$e = \{f_{t_1}, \ldots, f_{t_{21}}\},$$

a set of 21 independent component experiments, where $t_1, \ldots, t_{21}$ are the known types of the 21 patients.

Suppose that a new patient, who is already known to have Cushing's syndrome but of as yet unknown type, has urinary excretion rates of 9.0 mg/24h of tetrahydrocortisone and of 1.50 mg/24h of pregnanetriol. On the basis of our past experience – the informative experiment *e* – and on the basis of the

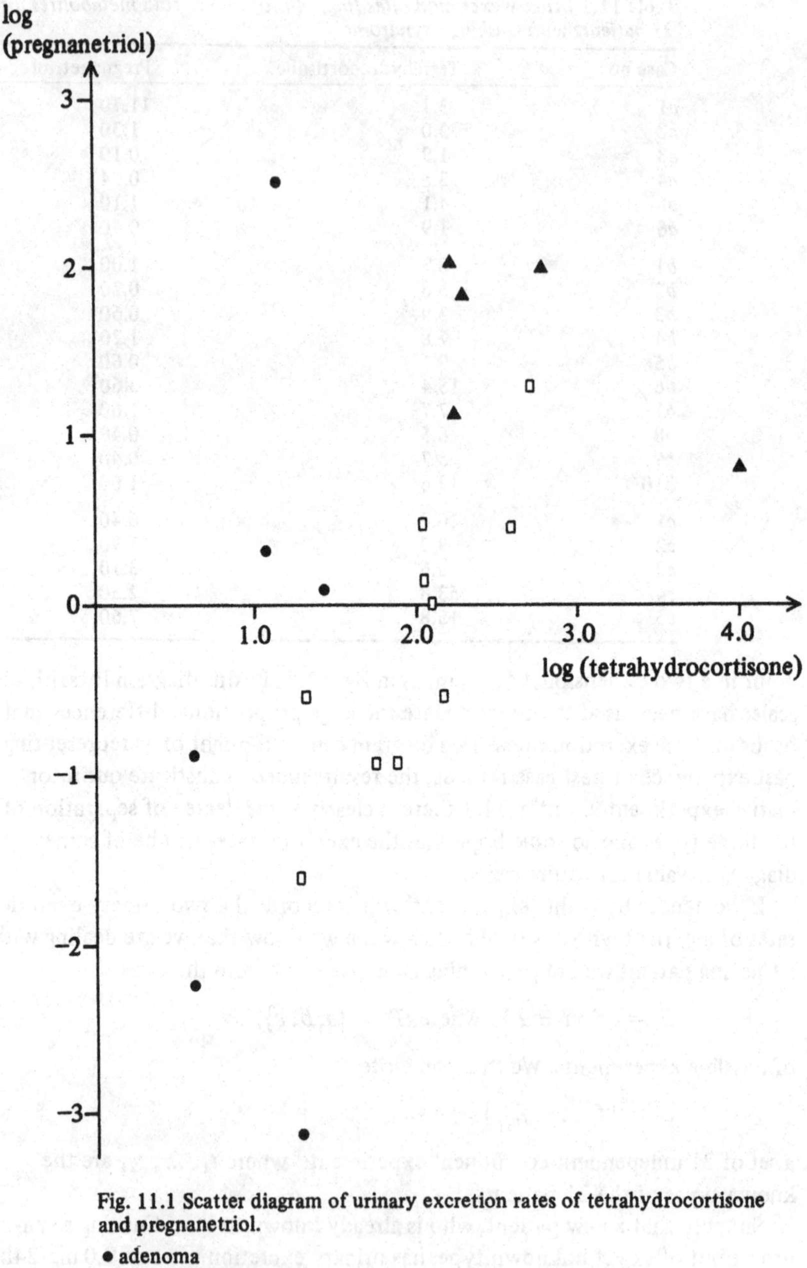

Fig. 11.1 Scatter diagram of urinary excretion rates of tetrahydrocortisone and pregnanetriol.

- ● adenoma
- ▫ bilateral hyperplasia
- ▲ carcinoma

outcome of these diagnostic tests on the new patient what plausibilities ought we to attach to the three types for this new patient? This is the nature of a diagnostic problem. If we denote the unknown type of the new patient by u then his results on the diagnostic tests constitute an outcome of a performance of f_u. We thus have a problem analogous to the calibration problem of chapter 10. We have already performed 'future' experiment f_u from the class F but we do not know whether u is a, b or c. The only difference between calibration and diagnosis lies in the fact that the set T is usually a continuum in a calibration problem whereas T is a finite set in a diagnostic problem.

We have described diagnosis as the assessment of the plausibilities of the various possible types rather than the definite allocation of the patients to one of the types. In other words we are regarding diagnosis more as a problem of inference than one of decision. We shall return to this point in §11.7, where we discuss the problem of diagnosis under a utility structure, and again in chapter 12, where diagnosis is seen as a kind of mental resting place in the search for a suitable treatment.

11.2 The diagnostic distribution

Just as in §10.2 we recognised the calibrative distribution as a sensible means of describing the plausibility of the possible indices of the new subject, so we shall be able to construct a diagnostic distribution which, for a new case, assigns plausibilities to each of the finite set of types. As in our treatment of calibration we again set down explicitly a set of assumptions as a basis for the development of statistical diagnosis. For many diagnostic situations only eight simple assumptions seem to be required, and acceptance of these has far-reaching consequences. We do not imply that all diagnostic problems conform to this pattern; our purpose is rather to emphasise the underlying assumptions so that it is easier to examine the relevance of the predictive method to a new situation and to pinpoint where any adjustment may have to be made.

For convenience of reference we first present compactly the eight assumptions, then discuss their interpretation and relevance, and finally follow through their consequences.

Assumptions

D1

Each case belongs to one and only one of a finite set $T = \{1, \dots, r\}$ of possible *types*.

D2

For each case a finite set $G = \{1, \dots, d\}$ of possible *features* may be observed. any observed *feature vector* falling within a given sample space $\mathbf{X}$.

D3
The class of probabilistic models considered as possible descriptions of the generation of *case records* (that is, types and associated feature vectors) is indexed by a finite-dimensional vector $(\boldsymbol{\Psi}, \boldsymbol{\theta})$ with $\boldsymbol{\Psi} \in \boldsymbol{\Psi}, \boldsymbol{\theta} \in \boldsymbol{\Theta}$.

In the following three assumptions $(t, \mathbf{x})$ with $t \in T, \mathbf{x} \in \mathbf{X}$ is a typical case record, and $\mathbf{z} = \{\mathbf{z}_1, \ldots, \mathbf{z}_n\}$ any given set of n case records.

D4
$p(t|\boldsymbol{\Psi}, \boldsymbol{\theta}) = p(t|\boldsymbol{\Psi}) \quad (t \in T)$ for every $\boldsymbol{\Psi} \in \boldsymbol{\Psi}, \boldsymbol{\theta} \in \boldsymbol{\Theta}$.

D5
$p(\mathbf{x}|t, \boldsymbol{\Psi}, \boldsymbol{\theta}) = p(\mathbf{x}|t, \boldsymbol{\theta}) \quad (\mathbf{x} \in \mathbf{X})$ for every $t \in T, \boldsymbol{\Psi} \in \boldsymbol{\Psi}, \boldsymbol{\theta} \in \boldsymbol{\Theta}$.

D6
$p(\mathbf{z}|\boldsymbol{\Psi}, \boldsymbol{\theta}) = \prod_{i=1}^{n} p(\mathbf{z}_i|\boldsymbol{\Psi}, \boldsymbol{\theta})$ for every set $\mathbf{z} = (\mathbf{z}_1, \ldots, \mathbf{z}_n)$

of case records.

D7
$p(\boldsymbol{\Psi}, \boldsymbol{\theta}) = p(\boldsymbol{\Psi})p(\boldsymbol{\theta}) \quad (\boldsymbol{\Psi} \in \boldsymbol{\Psi}, \boldsymbol{\theta} \in \boldsymbol{\Theta})$.

D8
For a case of unknown type u

$p(u|\boldsymbol{\Psi}) = \psi_u \; (u = 1, \ldots, d)$,

where ψ_u is the uth component of $\boldsymbol{\Psi}$, and

$\boldsymbol{\Psi} = \{\boldsymbol{\Psi}: \psi_t \geqslant 0, \quad \Sigma_T \psi_t = 1\}$

is the d-dimensional simplex.

There are two aspects of assumption D1 – the exhaustiveness and exclusiveness of the categories of T. The first aspect asserts that a case arriving for diagnosis by a system based on T does fall into one of the types of T. In example 11.1 if we adopt $T = \{a, b, c\}$ we have then no direct means of categorising say a patient who displays all the symptoms of Cushing's syndrome but who turns out to have a fourth type d of the disease. If we are to use this assumption wisely therefore we will choose as sets of types for consideration only those which have this property of exhaustiveness. Even for a set T chosen for this property we would still be wise to allow for the possibility of misdirected cases. We shall see in §11.4 that we can to some extent monitor for this possibility of wrong referral. The second aspect of exclusiveness asserts that a case cannot belong to more than one type, or, in medical terminology, have a dual or multiple pathology. If in example 11.1 it were possible for a patient to have both an adenoma and bilateral hyperplasia then to meet the assumption we would have to consider four categories (*a*) adenoma only, (*b*) hyperplasia only,

(c) carcinoma, ($d = ab$) adenoma and hyperplasia.

Assumption D2 simply asserts that for each case there are certain features (for example, (1) urinary excretion rate of tetrahydrocortisone, (2) diastolic blood pressure, (3) sex, and so on) which may be observed. The sample space **X** in a medical diagnostic situation may be large.

Assumption D3 expresses the usual statistical hope that we can describe the variability we observe in the case records in terms of some family of distributions indexed by a finite parameter. The partition of the parameter into $\boldsymbol{\Psi}$ and $\boldsymbol{\theta}$ is given meaning by the following two assumptions.

Assumption D4 asserts that, as far as the probabilistic mechanism which determines the type of a case is concerned, if we know $\boldsymbol{\Psi}$ then we need not know $\boldsymbol{\theta}$. In other words the parameter $\boldsymbol{\Psi}$ characterises the arrival pattern of types and so we shall term it the *arrival parameter*.

For diagnosis to be feasible we must hope that the way in which feature vectors arise depends to some extent on the type of the case. Assumption D5 states that this conditional distribution of feature vector given the type does not depend on the arrival parameter. In other words $\boldsymbol{\theta}$ is that aspect of the parameter which characterises the dependence of feature vector on type. For this reason we refer to $\boldsymbol{\theta}$ as the *structural parameter*. If in example 11.1 we were to make the assumption that the three distributions of feature vectors associated with the three types are bivariate normal with mean vectors $\boldsymbol{\mu}_1, \boldsymbol{\mu}_2, \boldsymbol{\mu}_3$ and covariance matrices $\boldsymbol{\Sigma}_1, \boldsymbol{\Sigma}_2, \boldsymbol{\Sigma}_3$ then the structural parameter $\boldsymbol{\theta}$ could be taken as $(\boldsymbol{\mu}_1, \boldsymbol{\Sigma}_1, \boldsymbol{\mu}_2, \boldsymbol{\Sigma}_2, \boldsymbol{\mu}_3, \boldsymbol{\Sigma}_3)$. In this case the distribution of feature vectors for the first type is $\text{No}_d(\boldsymbol{\mu}_1, \boldsymbol{\Sigma}_1^{-1})$ and so $p(\mathbf{x} \mid t = a, \boldsymbol{\theta})$ could be reduced to $p(\mathbf{x} \mid t = a, \boldsymbol{\mu}_1, \boldsymbol{\Sigma}_1)$, but the form of the theoretical development is kept clearer if we retain the complete $\boldsymbol{\theta}$ in the parametrisation of each of the feature vector distributions.

Assumption D6 states that, for given arrival and structural parameters, case records are statistically independent. While this is a sufficiently realistic assumption in a great number of diagnostic problems it would require careful reconsideration if any of the types were contagious or had a substantial genetic character. For example, if in two case records $(t_1, \mathbf{x}_1)$, $(t_2, \mathbf{x}_2)$ the vectors $\mathbf{x}_1$ and $\mathbf{x}_2$ provide evidence of a family relationship then in such a situation there might be an implied dependence between t_1 and t_2.

Assumption D7 states that prior to investigation of past case records any information that we have concerning the arrival pattern is independent of any information we have concerning the feature structure of the types. We shall see that the prior independence postulated in this assumption persists in a posterior form when the data on which updating is based consists of case records, which by definition are associated with cases of known types. In its posterior form it appears to be a tacit assumption of most statistical diagnostic methods. We emphasise that at this point of the formulation we are making no other particular assumptions about the form of $p(\boldsymbol{\psi})$ and $p(\boldsymbol{\theta})$.

The nature of the arrival parameter is clarified by assumption D8. For a given arrival pattern vector $\boldsymbol{\Psi}$ the r components ψ_u $(u = 1, \ldots, r)$ are the probabilities we immediately associate with the r possible types without any additional evidence. The vector $\boldsymbol{\Psi}$ is thus a probability vector, with elements summing to unity, which justifies the restriction to the r-dimensional simplex $\boldsymbol{\Psi}$.

In the analysis of calibration problems we had to distinguish between natural and designed calibration experiments, and we require to make the same distinction here. To achieve this we can consider assumption D4 in greater detail. As it stands this assumption does not specify the way in which the probabilistic mechanism determining the type of a past case depends on $\boldsymbol{\Psi}$. If the past case records constituting the informative experiment e have arisen naturally with the same arrival pattern as is anticipated for new cases then we would make the further assumption that

$$p(t\,|\,\boldsymbol{\Psi}) = \psi_t \tag{11.1}$$

for a past case record $(t, \mathbf{x})$. In some circumstances such a 'natural' informative experiment may be difficult to achieve. For example, if the incidence of one of the types is small then in order to obtain sufficient information on the feature structure for that type it may be necessary to seek the referral of such types from sources other than those immediately available. For such selected past case records we have a 'designed' informative experiment, and we can recognise this feature in assumption D4 by making the further assumption that $p(t\,|\,\boldsymbol{\Psi})$ does not depend on $\boldsymbol{\Psi}$:

$$p(t\,|\,\boldsymbol{\Psi}) = p(t). \tag{11.2}$$

We can thus retain all our assumptions while recognising that D4 simply asserts a possible dependence on $\boldsymbol{\Psi}$. For a natural informative experiment (11.1) holds and the informative experiment will clearly alter our prior assertion $p(\boldsymbol{\psi})$. For a designed informative experiment (11.2) holds and our prior distribution $p(\boldsymbol{\Psi})$ will be unaltered by such an experiment.

We now suppose that we have the case records $\mathbf{z} = \{\mathbf{z}_1, \ldots, \mathbf{z}_n\}$, where $\mathbf{z}_i = (t_i, \mathbf{x}_i)$, of n diagnosed cases, and the feature vector $\mathbf{y}$ of an undiagnosed case of unknown type u. Thus the complete parameter is $(\boldsymbol{\psi}, \boldsymbol{\theta}, u)$, the data are $(\mathbf{y}, \mathbf{z})$ and the general problem to resolve is how a prior distribution $p(\boldsymbol{\psi}, \boldsymbol{\theta}, u)$ on the parameter set is transformed to a posterior distribution $p(\boldsymbol{\psi}, \boldsymbol{\theta}, u\,|\,\mathbf{y}, \mathbf{z})$ on the basis of the data $(\mathbf{y}, \mathbf{z})$ and the set D of assumptions D1–D8. The change is of course brought about by an application of Bayes's theorem with likelihood $p(\mathbf{y}, \mathbf{z}\,|\,\boldsymbol{\Psi}, \boldsymbol{\theta}, u)$, and so to obtain insight into the process we must first discover what the implications of D are for the prior distribution and the likelihood. The structure of the prior is readily obtained:

$$
\begin{aligned}
p(\Psi,\Theta,u) &= p(\Psi,\Theta)p(u|\Psi,\Theta)\\
&= p(\Psi,\Theta)p(u|\Psi) \text{ by D4}\\
&= p(\Psi)p(\Theta)p(u|\Psi) \text{ by D7.} \qquad (11.3)
\end{aligned}
$$

To obtain the likelihood we note that

$$
\begin{aligned}
p(\mathbf{y},\mathbf{z}|\Psi,\Theta,u) &= p(u,\mathbf{y};\mathbf{z}|\Psi,\Theta)/p(u|\Psi,\Theta)\\
&= \frac{p(u,\mathbf{y}|\Psi,\Theta)\prod_{i=1}^{n} p(\mathbf{z}_i|\Psi,\Theta)}{p(u|\Psi,\Theta)} \text{ by D6}\\
&= p(\mathbf{y}|u,\Psi,\Theta)\prod_{i=1}^{n} p(t_i|\Psi,\Theta)p(\mathbf{x}_i|t_i,\Psi,\Theta)\\
&= p(\mathbf{y}|u,\Theta)\prod_{i=1}^{n} p(t_i|\Psi)\prod_{i=1}^{n} p(\mathbf{x}_i|t_i,\Theta) \text{ by D4 and D5}\\
&= p(\mathbf{y}|u,\Theta)p(\mathbf{t}|\Psi)p(\mathbf{x}|\mathbf{t},\Theta) \qquad (11.4)
\end{aligned}
$$

in an obvious shortened notation, where $\mathbf{t} = (t_1, \ldots, t_n)$, $\mathbf{x} = (\mathbf{x}_1, \ldots, \mathbf{x}_n)$.

Then, by Bayes's Theorem and using the symbol $\propto$ to indicate that the factor of proportionality does not depend on Ψ, Θ or u, we have

$$
\begin{aligned}
p(\Psi,\Theta,u|\mathbf{y},\mathbf{z}) &\propto p(\Psi)p(\Theta)p(u|\Psi)p(\mathbf{y}|u,\Theta)p(\mathbf{t}|\Psi)p(\mathbf{x}|\mathbf{t},\Theta)\\
&\propto p(\Psi|\mathbf{t})p(u|\Psi)p(\Theta|\mathbf{z})p(\mathbf{y}|u,\Theta), \qquad (11.5)
\end{aligned}
$$

where

$$p(\Psi|\mathbf{t}) = p(\Psi)p(\mathbf{t}|\Psi)\Big/\int_{\Psi} p(\Psi)p(\mathbf{t}|\Psi)d\Psi, \qquad (11.6)$$

$$
\begin{aligned}
p(\Theta|\mathbf{z}) &= p(\Theta|\mathbf{t},\mathbf{x})\\
&= p(\Theta)p(\mathbf{x}|\mathbf{t},\Theta)\Big/\int_{\Theta} p(\Theta)p(\mathbf{x}|\mathbf{t},\Theta)d\Theta. \qquad (11.7)
\end{aligned}
$$

If we then write

$$p(u|\mathbf{t}) = \int_{\Psi} p(u|\Psi)p(\Psi|\mathbf{t})d\Psi, \qquad (11.8)$$

$$p(\mathbf{y}|u,\mathbf{z}) = \int_{\Theta} p(\mathbf{y}|u,\Theta)p(\Theta|\mathbf{z})d\Theta, \qquad (11.9)$$

we can express the final general result in the following form:

$$p(\Psi,\Theta,u|\mathbf{y},\mathbf{z},D) = \frac{p(\Psi|\mathbf{t})p(u|\Psi)p(\Theta|\mathbf{z})p(\mathbf{y}|u,\Theta)}{\sum_{u\in T} p(u|\mathbf{t})p(\mathbf{y}|u,\mathbf{z})} \qquad (11.10)$$

where the inclusion of the conditioning D is to emphasise the dependence of the form on the assumptions. The full generality of this result would be required for such problems as updating the diagnostic method on the basis of new cases of unconfirmed type (Aitchison, Habbema and Kay, 1975); for our immediate diagnostic objective we are interested in the marginal distribution for u:

$$p(u\,|\,\mathbf{y}, \mathbf{z}) = \frac{p(u\,|\,\mathbf{t})p(\mathbf{y}\,|\,u, \mathbf{z})}{\sum\limits_{u \in T} p(u\,|\,\mathbf{t})p(\mathbf{y}\,|\,u, \mathbf{z})}, \tag{11.11}$$

the density function of the *diagnostic distribution.*

For a natural informative experiment an interpretation of (11.11) is that it is simply the conversion of an assessment $p(u\,|\,\mathbf{t})$ after the types $\mathbf{t}$ of the past cases are known, but prior to any information concerning the $n+1$ feature vectors $\mathbf{y}, \mathbf{z}$, to a posterior assessment $p(u\,|\,\mathbf{y}, \mathbf{t}, \mathbf{x})$ by way of Bayes's theorem and with $p(\mathbf{y}\,|\,u, \mathbf{t}, \mathbf{x})$ or $p(\mathbf{y}\,|\,u, \mathbf{z})$ playing the role of the likelihood function. Special interest then centres on $p(u\,|\,\mathbf{t})$ and $p(\mathbf{y}\,|\,u, \mathbf{z})$.

First we note that, from (11.8) and D8,

$$\begin{aligned} p(u\,|\,\mathbf{t}) &= \int_{\boldsymbol{\Psi}} \psi_u p(\boldsymbol{\Psi}\,|\,\mathbf{t}) d\boldsymbol{\Psi} \\ &= \mathrm{E}(\psi_u\,|\,\mathbf{t}). \end{aligned} \tag{11.12}$$

Hence in so far as inference concerning the category of the new case is concerned uncertainty about $\boldsymbol{\Psi}$ is involved only in the form $\mathrm{E}(\psi_u\,|\,\mathbf{t})$. We do not have to be able to provide a complete picture of the uncertainty in $p(\boldsymbol{\Psi}\,|\,\mathbf{t})$ but only the mean vector $\mathrm{E}(\boldsymbol{\Psi}\,|\,\mathbf{t})$ of this distribution.

The distribution $p(\mathbf{y}\,|\,u, \mathbf{z})$ defined by (11.9) is the now familiar predictive distribution. For a new case in known category u it provides, on the basis of prior information $p(\boldsymbol{\theta})$, the past records $\mathbf{z}$ and the assumptions D, an assessment of the probabilities of the possible feature vectors $\mathbf{y}$ we may observe on the case.

For a designed informative experiment the consequences of (11.2), (11.6) and (11.8) are that

$$p(u\,|\,\mathbf{t}) = p(u), \tag{11.13}$$

and we re-emphasise that the specification of $p(u)$ must then be based on sources other than the informative experiment.

We shall apply the predictive diagnostic method expressed by (11.11) to the case where the distributions of feature vectors for each given type are multinormal, say $\mathrm{No}_d(\boldsymbol{\mu}_t, \boldsymbol{\tau}_t)$ for type t. Suppose that of the n past case records n_t are of type $t\,(t = 1, \ldots, r)$ so that $n_1 + \ldots + n_r = n$. For the later application we record here the appropriate predictive distributions for vague

$NoWi_d$ priors, and must distinguish between two situations. The first is where we make no assumption about the equality of the covariance matrices of the r feature distributions. If $\bar{\mathbf{x}}_t$ and $\mathbf{S}_t$ denote the vector mean and the covariance matrix for the n_t feature vectors of type t then $\mathbf{m}_t = \bar{\mathbf{x}}_t$ and $\mathbf{v}_t = (n_t - 1)\mathbf{S}_t$ are independently distributed as $No_d(\boldsymbol{\mu}_t, n_t\boldsymbol{\tau}_t)$ and $Wi_d(n_t - 1, \boldsymbol{\tau}_t)$. We can then apply case 6 of table 2.3 to obtain

$$p(\mathbf{y}\,|\,u, \mathbf{z}) = St_d\left\{n_u - 1, \mathbf{m}_u, \left(1 + \frac{1}{n_u}\right)\frac{\mathbf{v}_u}{n_u - 1}\right\}. \tag{11.14}$$

The second situation is where we make the assumption that the covariance matrices of the r feature distributions are equal, say $\boldsymbol{\tau}_t = \boldsymbol{\tau}\,(t \in T)$. If $\mathbf{S}$ is the pooled covariance matrix of r different sets of observed feature vectors,

$$\mathbf{S} = \sum_{t \in T} (n_t - 1)\,\mathbf{S}_t/(n - r), \tag{11.15}$$

then $\mathbf{v} = (n - r)\mathbf{S}$ is distributed as $Wi_d(n - r, \boldsymbol{\tau})$ independently of $\mathbf{m}_t = \bar{\mathbf{x}}_t$ $(t \in T)$. Then again by case 6 of table 2.3 we have

$$p(\mathbf{y}\,|\,u, \mathbf{z}) = St_d\left\{n - r, \mathbf{m}_u, \left(1 + \frac{1}{n_u}\right)\frac{\mathbf{v}}{n - r}\right\}. \tag{11.16}$$

We also record the easily verifiable fact that if for a natural informative experiment we adopt for $p(\boldsymbol{\Psi})$ the vague Dirichlet distribution $Di(\mathbf{g}, h)$ with $\mathbf{g} \to \mathbf{0}, h \to 0$ then

$$p(u\,|\,\mathbf{t}) = \frac{n_u}{n}, \tag{11.17}$$

the proportion of type u cases in the past case records.

11.3 An illustrative example

We now illustrate the use of the diagnostic distribution by applying the results of §11.2 to example 11.1. We make the assumption that the logarithms of the excretion rates are bivariate normal for each type and record the required summary of the data of table 11.1, first transformed by taking natural logarithms:

$$n_1 = 6, \quad n_2 = 10, \quad n_3 = 5; \quad r = 3, \quad d = 2;$$

$$\bar{\mathbf{x}}_1 = \begin{bmatrix} 1.0433 \\ -0.6034 \end{bmatrix}, \quad \bar{\mathbf{x}}_2 = \begin{bmatrix} 2.0073 \\ -0.2060 \end{bmatrix}, \quad \bar{\mathbf{x}}_3 = \begin{bmatrix} 2.7097 \\ 1.5998 \end{bmatrix};$$

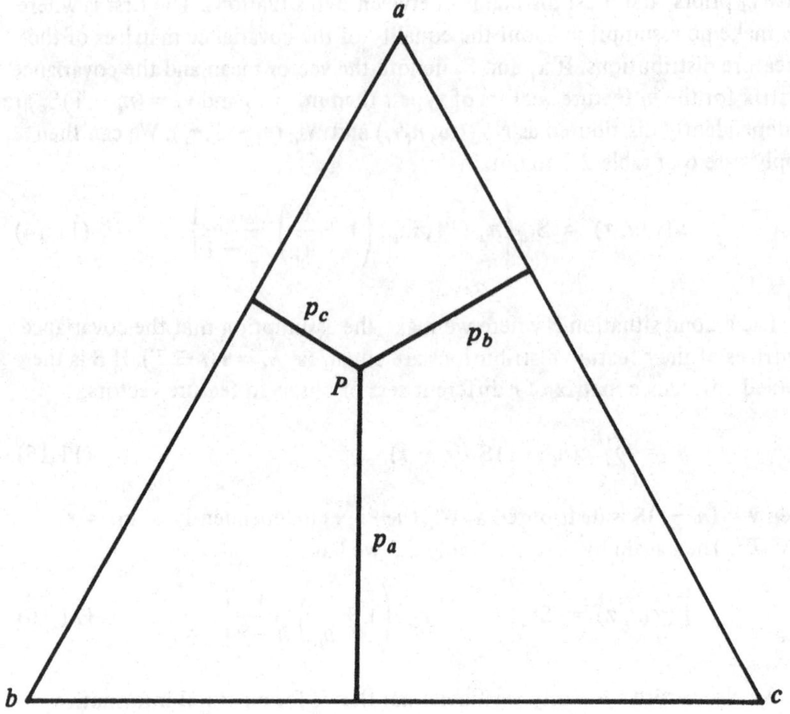

Fig. 11.2 Reference triangle for plausibility assessments about Cushing's syndrome.

$$\mathbf{S}_1 = \begin{bmatrix} 0.11069 & 0.12389 \\ 0.12389 & 4.08910 \end{bmatrix}, \quad \mathbf{S}_2 = \begin{bmatrix} 0.21187 & 0.32413 \\ 0.32413 & 0.72030 \end{bmatrix},$$

$$\mathbf{S}_3 = \begin{bmatrix} 0.55522 & -0.24224 \\ -0.24224 & 0.28850 \end{bmatrix}, \quad \mathbf{S} = \begin{bmatrix} 0.26006 & 0.14265 \\ 0.14265 & 1.56012 \end{bmatrix}.$$

There is some evidence in $\mathbf{S}_1, \mathbf{S}_2, \mathbf{S}_3$ that it would be unreasonable to adopt the assumption that $\boldsymbol{\tau}_1 = \boldsymbol{\tau}_2 = \boldsymbol{\tau}_3$ so that we shall first use form (11.14) in our diagnostic assessment. The data in table 11.1 did not in fact arise from a natural informative experiment and so we shall quote results for the case where $p(u) = 1/3$ for $u = 1, 2, 3$. The diagnostic distribution for other $p(u)$ is easily derived by a simple weighting of the quoted diagnostic distribution.

To present the results of the application of (11.11) we use a simple diagrammatic representation. In fig. 11.2 the equilateral triangle *abc* with unit altitude is such that for any point *P* within it the sum of the perpendiculars

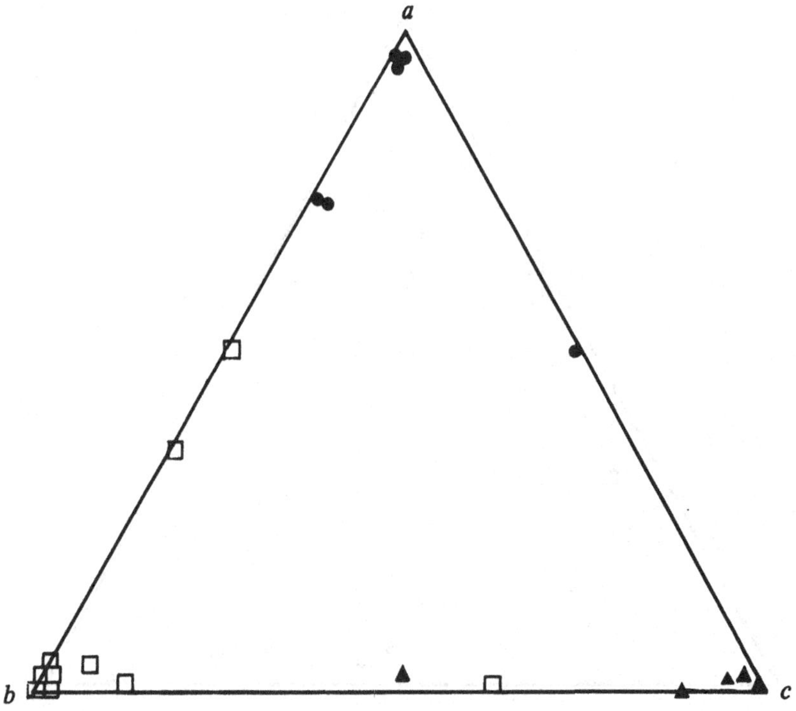

Fig. 11.3 Diagnostic assessments on the basis of unequal covariance matrices for the 21 past cases of Cushing's syndrome.

- ● adenoma
- □ bilateral hyperplasia
- ▲ carcinoma

p_a, p_b, p_c is unity. Thus any statement which places plausibilities p_a, p_b, p_c on the three types a, b, c can be represented by a unique point $P(p_a, p_b, p_c)$ in the triangle abc; and conversely each point in the triangle represents a unique plausibility statement about a, b, c. Roughly speaking with this representation the nearer a point is to a vertex the more plausible the type associated with that vertex is being regarded. The distribution (11.11) associated with any feature vector y thus gives a point within triangle abc. Application of this diagnostic method to each of the feature vectors of table 11.1 produces 21 such points (fig. 11.3) and some measure of the potential of the method is given by the extent to which these points are close to the correct vertex.

For comparison purposes we show in fig. 11.4 the corresponding points which arise if the assumption of equal covariance matrices is adopted. As might have been anticipated in this particular example recognition of the

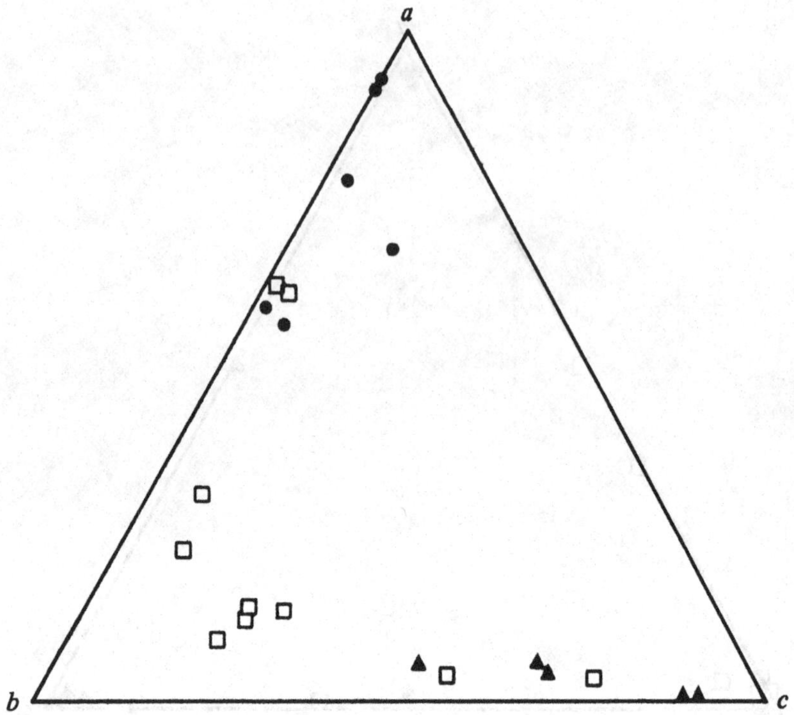

Fig. 11.4 Diagnostic assessments on the basis of equal covariance matrices for the 21 past cases of Cushing's syndrome.

● adenoma
□ bilateral hyperplasia
▲ carcinoma

evidence of unequal covariance matrices leads on the whole to a firmer diagnostic view.

Fig. 11.5 shows the diagnostic distributions based on (11.14) for six new cases listed in table 11.2. We shall return to a discussion of these cases in §11.4.

11.4 Monitoring for atypicality

In discussing the aspect of assumption D1 concerning the exhaustiveness of the set T of possible types we indicated that some form of monitoring is desirable to ensure that the decision to stream a new case into the speciality characterised by T has not been unreasonable. We have already met the basic

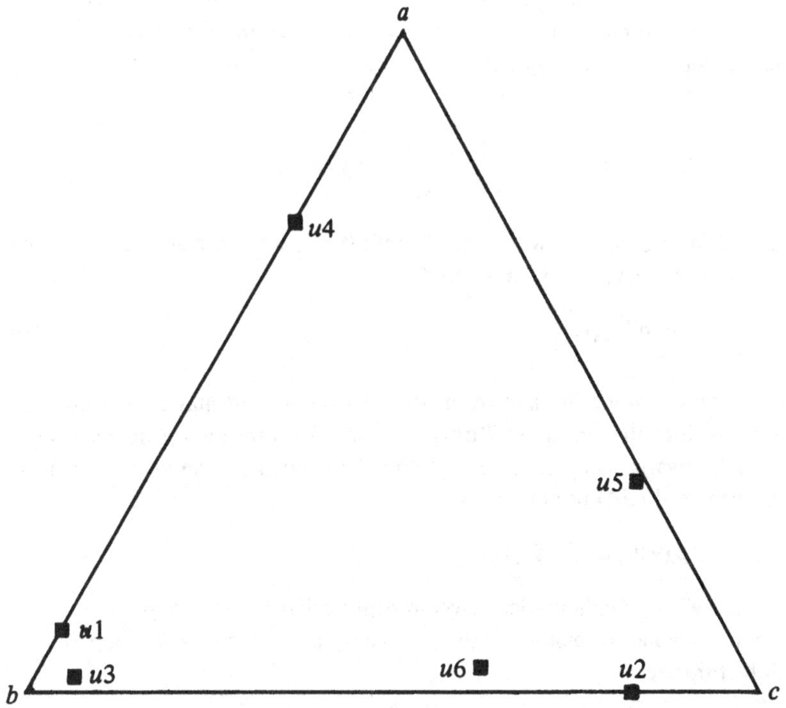

Fig. 11.5 Diagnostic assessments on the basis of unequal covariance matrices of 6 new cases of Cushing's syndrome.

Table 11.2 *Urinary excretion rates (mg/24h) of two steroid metabolites for 6 undiagnosed patients with Cushing's syndrome*

Case no.	Tetrahydrocortisone	Pregnanetriol
*u*1	5.1	0.4
*u*2	12.9	5.0
*u*3	13.0	0.8
*u*4	2.6	0.1
*u*5	30.0	0.1
*u*6	20.5	0.8

idea of monitoring in the discussion of the notion of 'past experience' in §4.3. A method of achieving a form of monitoring takes the following form. On the basis of past experience **z** of a particular type u the probability distribution associated with the feature vector **y** of any new patient is simply the predictive distribution $p(\mathbf{y} \mid u, \mathbf{z})$. Suppose that we are concerned about how typical of disease category u a patient with observed feature vector $\mathbf{y}_0$ is? We can regard any case with feature vector **y** for which $p(\mathbf{y} \mid u, \mathbf{z}) \geqslant p(\mathbf{y}_0 \mid u, \mathbf{z})$ as more typical of u than patient $\mathbf{y}_0$, or equivalently $\mathbf{y}_0$ as less typical than $\mathbf{y}$. We can thus construct for patient $\mathbf{y}_0$ a sensible index $\mathcal{I}_u(\mathbf{y}_0)$ of atypicality relative to disease category u as the probability (on the basis of past experience) that another patient is more typical than him. Thus

$$\mathcal{I}_u(\mathbf{y}_0) = \int_{\{y:p(y|u,z) \geq p(y_0|u,z)\}} p(\mathbf{y}\,|\,u, \mathbf{z})\,d\mathbf{y}. \tag{11.18}$$

Thus $\mathcal{I}_u$ is measured on the scale (0, 1) with 0 indicating the absolutely typical and 1 complete atypicality. If we find

$$\min_{u \in T} \mathcal{I}_u(\mathbf{y}_0) \tag{11.19}$$

near 1 then we would be right to suspect that the patient may have been channelled into the wrong set T of types, and take some appropriate action.

In a multivariate normal setting the predictive distribution takes a Student form such as (11.14) or (11.16), say

$$p(\mathbf{y}\,|\,u, \mathbf{z}) = \mathrm{St}_d(k, \mathbf{b}, \mathbf{c}).$$

Then following a mathematical development similar to that of §4.3 we arrive at a simple expression for the atypicality index for type u of a patient with feature vector $\mathbf{y}$:

$$\mathcal{I}_u(\mathbf{y}) = I_{q(\mathbf{y})/[q(\mathbf{y})+k]}\,\{\tfrac{1}{2}d, \tfrac{1}{2}(k-d+1)\} \tag{11.20}$$

where

$$q(\mathbf{y}) = (\mathbf{y}-\mathbf{b})'\mathbf{c}^{-1}(\mathbf{y}-\mathbf{b}). \tag{11.21}$$

For the 21 past cases of Cushing's syndrome the most atypical for each of the three types are as follows:

(*a*) Case $a4$ with atypicality $\mathcal{I}_a = 0.56$,
(*b*) Case $b3$ with atypicality $\mathcal{I}_b = 0.75$,
(*c*) Case $c4$ with atypicality $\mathcal{I}_c = 0.52$.

Table 11.3 shows the atypicality indices $\mathcal{I}_a, \mathcal{I}_b$ and $\mathcal{I}_c$ for the six new cases of Cushing's syndrome presented in table 11.2, together with the confirmed histopathological type after operation, where appropriate. For cases $u1$ to $u4$ these histopathological types are all in agreement with the predictive diagnostic assessment of fig. 11.5. For case $u5$ all three atypicality indices are large and we would be hesitant to regard the diagnostic distribution as expressing an appropriate view for this case. Rather we would draw attention to the possibility that the case has been wrongly referred to the set T of types. In fact the data for $u5$ arose from an irregularity in collection of the urine specimen, and is

Table 11.3 *Atypicality indices for the six new cases of Cushing's syndrome and the actual type confirmed by histopathology*

Case no.	$\mathcal{I}_a$	$\mathcal{I}_b$	$\mathcal{I}_c$	Confirmed type
$u1$	0.60	0.25	0.975	b†
$u2$	0.95	0.84	0.02	c
$u3$	0.95	0.80	0.91	b
$u4$	0.21	0.86	0.993	a
$u5$	0.991	0.9999	0.984	Irregular urine collection†
$u6$	0.981	0.978	0.89	d†

† See text.

presented here to show that atypicality indices can serve as a useful safeguard, particularly in higher-dimensional problems where the simplicity of fig. 11.1 is not available. Case $u6$ is actually of a type d of Cushing's syndrome not included in the set T. It is presented here as a sharp reminder that while monitoring for atypicality is a necessary discipline of any sensible diagnostic system it is not sufficient to guarantee that a new case outside T will not be differentially diagnosed within T in an apparently satisfactory way. For a full diagnostic analysis of Cushing's syndrome it is necessary to include type d in T and to extend the dimensionality of the feature vector from 2 to 15.

For any new case, in addition to reporting on atypicality we can also state whether or not it is outside previous experience of any particular type, in the sense of §4.3. In terms of table 11.3, for example, a new case is outside previous experience of type b if its atypicality is greater than the case of b of greatest atypicality in the basic set. Thus while case $u3$ is being diagnosed correctly as of type b it is outside previous experience of type b. This is not surprising since our previous experience of type b is small, being confined to ten case records. Indeed we can indicate how likely we are to see a new case of type b which falls outside this previous experience because this is simply assessed by

$$1 - \max\{\mathcal{I}_b(\mathbf{x}_i) : i = 1, \ldots, n_b\}, \tag{11.22}$$

where $\mathbf{x}_1, \ldots, \mathbf{x}_{n_b}$ are the feature vectors of the n_b cases of b in the basic set. For type b there is thus a probability of 0.25 of obtaining a new case outside previous experience. For types a and c the corresponding probabilities are 0.44 and 0.48.

11.5 Estimative and predictive diagnosis

A few simple assumptions have led us inevitably to a particular form of diagnostic assessment, in terms of the *predictive* diagnostic distribution. The statistical diagnostic methods currently widely advocated are of estimative type and are generally supported by an argument of the following kind. If we knew the structural parameter $\boldsymbol{\theta}$ then we could easily arrive at appropriate plausibilities for the unknown disease category u of a new patient with feature vector $\mathbf{y}$ from prior plausibilities $p(u)$, or possibly $p(u\,|\,\mathbf{t})$. We would simply apply Bayes's theorem in the form

$$p(u\,|\,\mathbf{y}) \propto p(u)p(\mathbf{y}\,|\,u,\boldsymbol{\theta}). \tag{11.23}$$

Recognising that we do not know $\boldsymbol{\theta}$ the *estimative* method replaces $\boldsymbol{\theta}$ by a suitable estimate $\hat{\boldsymbol{\theta}}(\mathbf{z})$, for example a maximum likelihood estimate, based on the past records $\mathbf{z}$. We could thus rewrite the estimative method as:

$$p(u\,|\,\mathbf{y},\mathbf{z}) \propto p(u)\,p(\mathbf{y}\,|\,u,\hat{\boldsymbol{\theta}}(\mathbf{z})). \tag{11.24}$$

We recall the form (11.11) of the predictive method,

$$p(u\,|\,\mathbf{y},\mathbf{z}) \propto p(u)\,p(\mathbf{y}\,|\,u,\mathbf{z}) \tag{11.25}$$

where

$$p(\mathbf{y}\,|\,u,\mathbf{z}) \propto \int_{\Theta} p(\mathbf{y}\,|\,u,\boldsymbol{\theta})p(\boldsymbol{\theta}\,|\,\mathbf{z})d\boldsymbol{\theta}. \tag{11.26}$$

Clearly (11.24) and (11.25) will be in good agreement if $p(\boldsymbol{\theta}|\mathbf{z})$ is highly concentrated at $\hat{\boldsymbol{\theta}}(\mathbf{z})$. While, by the usual large sample arguments, this will be the case when there is a substantial past experience, there are many areas of medicine where past experience is modest, and these are situations where data interpretation is at a premium. One way of interpreting the fallacy of the estimative method is that it takes no account of the sampling variability of the estimator $\hat{\boldsymbol{\theta}}(\mathbf{z})$ of $\boldsymbol{\theta}$. Whereas, through (11.26), the predictive method weights the possible distributions $p(\mathbf{y}\,|\,u,\boldsymbol{\theta})$ according to the plausibilities of the various $\boldsymbol{\theta}$.

A common dissatisfaction with the estimative method is that far too optimistic a picture is painted when the method is assessed on the basic set of patients and all estimates of the 'misclassification rates' for future patients are far too low. While this is to some extent attributable to such factors as exclusion of difficult cases from the set $\mathbf{z}$ of past records, undoubtedly one contributory factor is the use of the estimative rather than the predictive approach. The predictive method has a tendency to damp down the over-optimism to a realistic degree. That there must be this tendency can be seen from very simple considerations. If $r = 2, d = 1$ and the feature distributions for the two categories are normal then the difference between (11.24) and (11.25) is that, whereas the distribution of y in (11.24) is being treated as normal, that of (11.25) is the corresponding Student distribution; see fig. 11.6. The alteration to the prior odds for the new patient by the estimative method is by the factor $AP/BP = 8.0$, whereas by the predictive method the factor is $A'P'/B'P' = 2.5$.

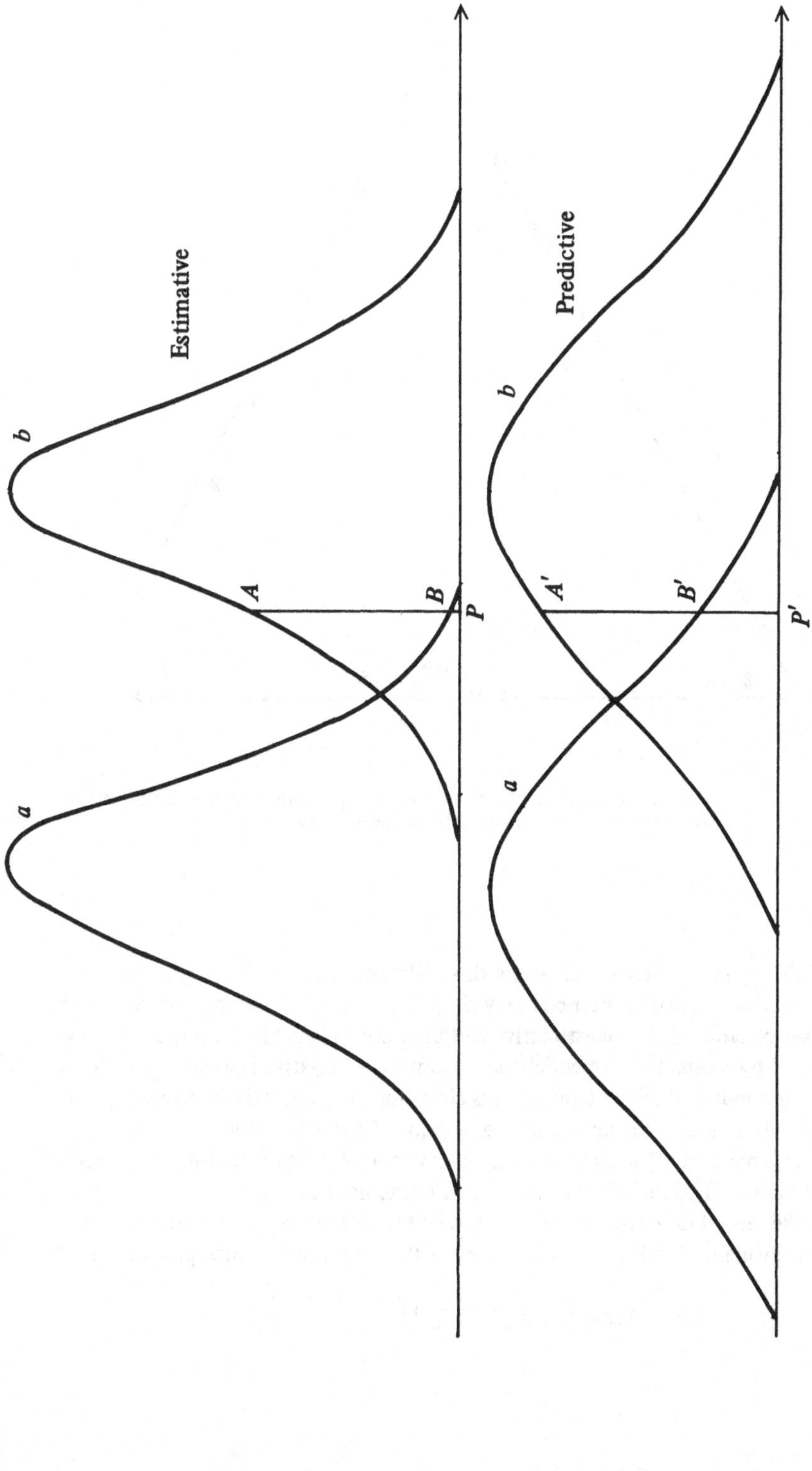

Fig. 11.6 Typical difference between estimative odds AP/BP and predictive odds $A'P'/B'P'$.

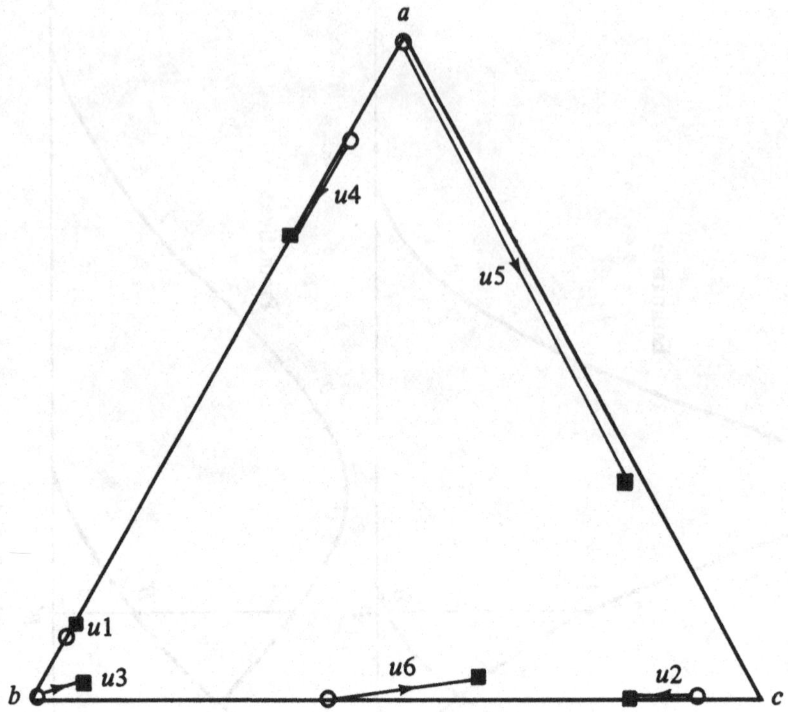

Fig. 11.7 Comparison between the estimative and predictive diagnostic assessments for the 6 new cases of Cushing's syndrome.
o estimative
■ predictive

For example 11.1 we illustrate this difference between the estimative and predictive diagnostic methods very simply. In fig. 11.7, for each of the six new cases of table 11.2, the estimative and predictive plausibility points are shown joined by a directed line which leads from the estimative assessment to the corresponding predictive point. It is clear that the estimative method tends on the whole to give the appearance of a firmer diagnostic view than the predictive method. That this tendency can be completely misleading in a practical problem will be seen in the application in the next section.

We record here for reference purposes the estimative counterpart of (11.20). For normally distributed feature vectors the estimative counterpart of (11.14) is

$$p(\mathbf{y}\,|\,u, \hat{\boldsymbol{\theta}}(\mathbf{z})) = \mathrm{No}_d(\bar{\mathbf{x}}_u, \mathbf{S}_u^{-1}) \tag{11.27}$$

and the estimative index of atypicality for type u of patient with feature vector $\mathbf{y}$ can be shown to be

$$\mathcal{I}_u(\mathbf{y}) = J_{Q(\mathbf{y})}(\tfrac{1}{2}d), \tag{11.28}$$

where

$$Q(\mathbf{y}) = (\mathbf{y}-\bar{\mathbf{x}}_u)'\mathbf{S}_u^{-1}(\mathbf{y}-\bar{\mathbf{x}}_u). \tag{11.29}$$

If the assumption of equal covariance matrices for the different types is made then $\mathbf{S}_u$ in (11.27) to (11.29) is simply replaced by $\mathbf{S}$, as defined in (11.15).

11.6 An application to differential diagnosis of Conn's syndrome

That the distinction we have made between estimative and predictive diagnosis is no mere hair-splitting is well illustrated by the application of these techniques to the differential diagnostic problem of Conn's syndrome posed as example 1.7. We shall not set out in detail the computations which are a straightforward application of the techniques of the preceding sections, but simply highlight the special aspects of this particular problem and report the widely differing results of the two techniques.

First we observe that there is appreciable skewness in some of the data; see in particular the aldosterone results for the adenoma patients. Some transformation is advisable to satisfy the normality assumption. In what follows we have used a blanket logarithmic transformation for each of the features; this appears to give reasonably symmetric distributions and so makes the normality assumption at least feasible. Secondly an assumption of equality of covariance matrices appears a doubtful proposition; for example, the aldosterone results for the bilateral hyperplasia patients are tightly packed compared with the wide spread of aldosterone results for the adenoma patients. We therefore work on the basis of different covariance matrices for the two types.

With these assumptions the structural parameter $\boldsymbol{\theta}$ consists of the two eight-dimensional mean vectors and the two covariance matrices of order 8, and so is an 88-dimensional parameter. The estimative method thus attempts to estimate this 88-dimensional parameter from 31 eight-dimensional vectors and to use each of the 88 component estimates as if it was the true value. Clearly such a method, by ignoring the obvious unreliability of the estimate, must run a risk of wild assessment. A first indication of the remarkable extent of this risk is seen in fig. 11.8 where we apply both the estimative and the predictive methods to the 31 patients in the basic set of table 1.6. Since there are just two types it is simplest to show the results in terms of the odds in favour of adenoma. We show these on the basis that the 31 case records form a natural, as opposed to a designed, informative experiment so that, by (11.17) we take

$$p(u = 1\,|\,\mathbf{t}) = \tfrac{20}{31}, \quad p(u = 2\,|\,\mathbf{t}) = \tfrac{11}{31}. \tag{11.30}$$

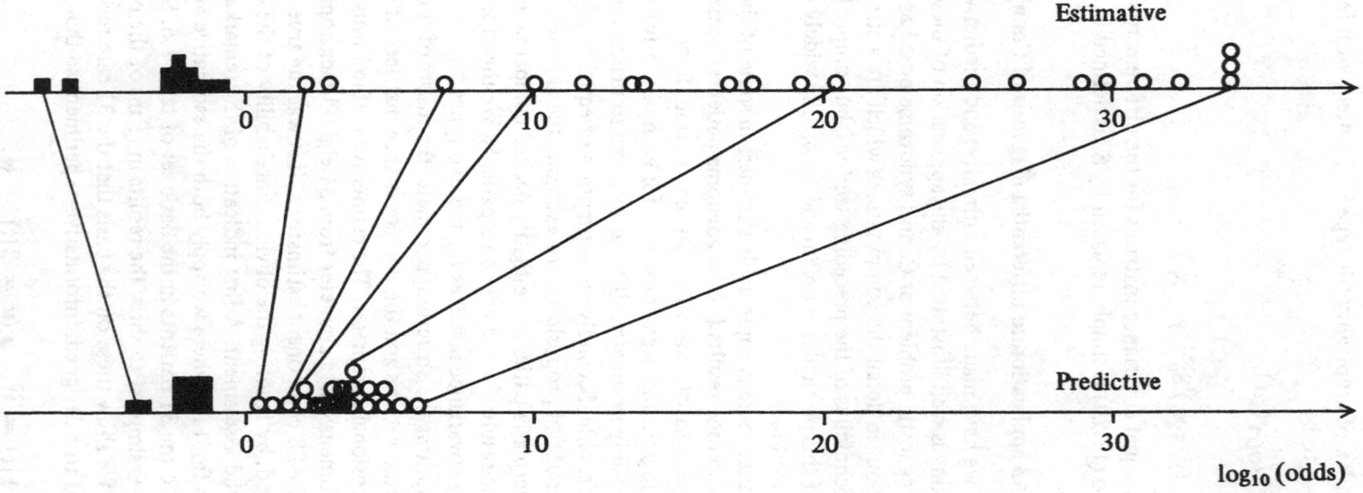

Fig. 11.8 Comparison of estimative and predictive odds on adenoma, shown on a $\log_{10}$ scale, for the 31 past cases of Conn's syndrome.

Table 11.4 *Feature vectors of four new cases of Conn's syndrome*

Case no.	Age (Years)	Concentrations in blood plasma: Na (meq/l)	K (meq/l)	CO_2 (meq/l)	Renin (meq/l)	Aldo-sterone (meq/l)	Blood Pressures: Systolic (mm Hg)	Diastolic (mm Hg)
*u*1	50	143.3	3.2	27.0	8.5	51.0	210	130
*u*2	49	142.6	2.3	36.0	6.2	35.7	192	125
*u*3	44	143.1	4.0	26.8	5.5	17.7	170	120
*u*4	53	142.8	4.0	26.3	3.6	65.8	260	145

Table 11.5 *Comparison of estimative and predictive odds and atypicality indices for four new cases of Conn's syndrome: syndrome*

Patient no.	Odds a/b		Atypicality indices			
			$u=a$		$u=b$	
	Estimative	Predictive	Estimative	Predictive	Estimative	Predictive
*u*1	$10^{12}/1$	64/1	0.69	0.32	1.00	0.76
*u*2	$10^{18}/1$	56/1	0.96	0.66	1.00	0.85
*u*3	1/245	1/80	0.90	0.54	0.85	0.14
*u*4	3700/1	1/2	0.98	0.71	1.00	0.60

(For a clinic anticipating a different incidence rate for the two types the appropriate modification of odds could be easily applied.) Because of the large proportional differences in the odds for the different cases the odds have been plotted on a logarithmic scale, $\log_{10}\{p(u=1\,|\,\mathbf{y},\mathbf{z})/p(u=2\,|\,\mathbf{y},\mathbf{z})\}$ being calculated by means of (11.11), (11.14) and (11.30).

From fig. 11.8 we see that the exaggerated confidence of assessments of odds of 10^{20} to 1 by the estimative method can be slashed to 10^3 to 1 by the predictive assessment. On the whole the estimative method gives much more extreme odds than the predictive method. It could be argued that we are re-applying the statistical diagnostic system to cases whose type we already know so that we might expect large odds to be assessed, so that perhaps the estimative method is the more realistic. But the differences with respect to new patients can be equally startling. Table 11.4 shows the feature vectors of four typical new patients known to have Conn's syndrome but of unknown type, case $u1$ being reproduced from table 1.6. The dramatic alterations of odds in some of these cases (table 11.5) is due to the fact that they are, in the technical sense of §4.3, outside the previous limited experience of both types, though not in any way atypical of the favoured disease. For such cases the predictive method acts with extra caution which is again very reasonable. Case $u4$ is particularly interesting; it is in fact a now-confirmed case of bilateral hyperplasia, not the clear case of adenoma assessed by the estimative method.

To sum up, the estimative method can give an exaggerated view of the evidence in favour of a particular type whereas the predictive method almost invariably moderates this view. This moderation is greatest when there is a limited past experience and in interpreting this we must remember that it has to be measured not in terms of the number of cases but in terms of the number of cases relative to the dimension of the feature vector. Eleven cases may be adequate experience if we are dealing with a one-dimensional feature, but they constitute a very limited experience for an eight-dimensional feature. We can see this in terms of the probabilities of obtaining a new patient within the previous experience of the two types. For adenoma this is 0.45, and for bilateral hyperplasia it is only 0.08.

11.7 Diagnosis under a utility structure

If diagnosis is to be regarded as a decision problem then the decision theory model must have the following components.

Parameter space. The set of unknown parameters can be identified with the index set T for the class F of possible feature determining experiments, the unknown type u of the case under consideration playing the role of the unknown state of nature.

Action set. An action a is simply a decision to act as if the type of the case were a, so that the action set A is the set T of possible types.

Utility function. To complete the specification of diagnosis as a decision problem we must therefore be in a position to attach a utility $U(a, u)$ to each possible action a in the face of each possible type u.

The decision problem is then to maximise the expected utility of having observed the past case records $\mathbf{z}$ and the feature vector $\mathbf{y}$ of the new case under consideration and of having taken action a. We have that

$$U(a, \mathbf{y}) = \sum_{u \in T} U(a, u) p(u | \mathbf{y}, \mathbf{z}), \tag{11.31}$$

where $p(u | \mathbf{y}, \mathbf{z})$ is the diagnostic distribution and where we have dropped from the notation the obvious dependence on $\mathbf{z}$. Thus if the diagnostic distribution has been evaluated there is no difficulty in evaluating $U(a, \mathbf{y})$ for each of the r possible actions a and in choosing the maximising action.

In medical diagnostic problems the difficulty lies in obtaining a realistic specification of the utility structure. Using utilities can be envisaged only in terms of the outcome of treatment allocated, with the almost overwhelming problem of expressing, for example, the advantage to a patient of some improvement in his condition in terms of the same utility units as the cost of treatment, not only in terms of monetary expense but also in terms of discomfort. Indeed only if there is a unique treatment associated with each type will it be sensible to formulate the diagnostic problem in terms of (11.31). Otherwise the problem is part of a more extensive sequential process and should be treated as such.

A popular device to by-pass this difficulty is to concentrate on minimising the overall misclassification or mistyping rate. In terms of a decision problem this is equivalent to adopting a utility structure

$$U(a,u) = \begin{cases} 1 & \text{if } a = u, \\ 0 & \text{if } a \neq u, \end{cases}$$

that is, placing zero utility on a mistyping of whatever kind, and unit utility to each correct typing. With this utility structure (11.31) becomes

$$U(a,\mathbf{y}) = p(a \mid \mathbf{y},\mathbf{z}),$$

so that we obtain the very simple rule: for a new patient with feature vector $\mathbf{y}$ take as decisive type the mode of the diagnostic distribution. While this simple rule has a strong intuitive appeal there is of course no guarantee that the real utility structure of the problem is such as to make this optimum.

History

The linear discriminant method of Fisher (1936) was the first statistical technique to be devised for the diagnostic type of problem considered in this chapter. A good account of this and other estimative methods is given in Anderson (1958), who incidentally presents though does not develop a method equivalent to the predictive method through a generalised likelihood ratio test argument. The predictive method is first presented explicitly in Geisser (1964) for the multivariate normal case, and in Dunsmore (1966). The great practical differences that there can be between the estimative and predictive methods appear to have been first noted by Aitchison and Kay (1974), who also consider the uses of indices of atypicality.

Problems

11.1 A diagnostic assessment between two psychiatric types, 1 and 2, of subject is to be made on the basis of a single binomial test on a subject who either responds or does not respond to the test stimulus. Of n naturally arising past subjects, firmly typed by other more elaborate means, n_1 and n_2 were of types 1 and 2. Suppose that the psychiatrist is persuaded to express his prior knowledge of the two response probabilities θ_1 and θ_2 as independent $\text{Be}(g_1, h_1)$, $\text{Be}(g_2, h_2)$ distributions. When the test was applied to the n past subjects x_i of the n_i of type i $(i = 1, 2)$ responded.

A new subject has just undergone the test and responded. How do you assess the plausibility of his being of type 1? Can you define indices of atypicality for this diagnostic problem?

11.2 Past experience **z** for a certain type consists of two observations x_1 and x_2 on a one-dimensional normally distributed feature. Show that the probability that a new case of this type has an observed feature falling between x_1 and x_2 is $2\Phi(1/\sqrt{2})-1=0.52$ on the estimative and $I_{1/4}(\frac{1}{2},\frac{1}{2})=\frac{1}{3}$ on the predictive assessment. Which do you regard as the more reasonable assessment, and why?

11.3 Construct a diagnostic system for Conn's syndrome (example 1.7) on the basis of the single feature K of table 1.6.

11.4 The 'disintegration times' of two types 1 and 2 of 'elementary particles' are assumed to be $\text{Ex}(\theta_1)$ and $\text{Ex}(\theta_2)$ with unknown indices θ_1, θ_2. In an experiment in which six naturally occurring particles were observed the following case records (particle type, disintegration time in ms) were recorded:

$$(1,47),\quad (2,75),\quad (1,17),\quad (1,32),\quad (2,31),\quad (1,19).$$

In two further independent experiments the disintegration times of two recorded particles of unknown type were 40 and 70 ms. Obtain the diagnostic distributions and atypicality indices for these two particles, both on a predictive and on an estimative basis.

How do you assess the probability that a new type 2 particle is outside previous experience?

11.5 For a problem of differential diagnosis of three types on the basis of two features the case records of 18 pathologically diagnosed patients are shown in the table below. Construct and compare estimative and predictive diagnostic systems on the basis of this past experience. Apply both systems to the new patients listed below.

Past case records

Case no.	Feature 1	2
*a*1	1.38	1.95
*a*2	0.67	1.73
*a*3	1.86	2.91
*a*4	1.12	2.42
*a*5	1.45	2.33
*b*1	2.80	1.40
*b*2	1.85	2.71
*b*3	2.12	2.00
*b*4	1.71	2.37
*b*5	2.40	1.48
*b*6	2.48	1.93
*c*1	2.32	2.43
*c*2	2.10	2.35
*c*3	2.08	2.62
*c*4	2.63	2.44
*c*5	2.35	2.85
*c*6	2.52	2.30
*c*7	2.56	2.75

New patients

Case no.	Feature 1	2
*u*1	2.30	1.95
*u*2	1.81	2.83
*u*3	1.44	1.38
*u*4	2.34	2.47
*u*5	3.00	2.32

11.6 Construct a diagnostic system for Conn's syndrome (example 1.7) based on the three plasma concentrations of the electrolytes Na, K and CO_2 as given in table 1.6.

11.7 Complete the analysis of problem 1.6.

12
Treatment allocation

12.1 The nature of a treatment allocation problem

When a treatment is applied to an object or individual it is with the express purpose of altering the future of that object or individual. Thus when we choose one of a number of possible refining processes for a batch of raw material we intend that the batch will in the future attain some desirable quality. When we select a method of machining an industrial component we have in mind some future characteristic of the component. When we prescribe a particular treatment for a patient we hope that some specific aspect of his future condition will be more agreeable than his present state of disease. Because of this preoccupation with the future state of an object or individual it will not be surprising to find that statistical prediction analysis has an important role to play in the problem of treatment allocation.

In the examples already mentioned there are three basic sets which must clearly play an important role. First we suppose that the present state or indicator t of the individual unit under consideration belongs to some specifiable set T of possible initial states or indicators. Secondly, there is some set A of possible treatments from which we have to select a treatment a to apply to the individual unit. Thirdly we must to some extent assess the effectiveness of treatment in terms of the future state or response y attained by the unit after application of treatment; we thus have to be in a position to envisage the set Y of possible future states or responses. In order to see the exact roles played by these sets and to obtain a clearer insight into the nature of treatment allocation problems we now consider specific examples.

Example 12.1

Quality improving process. An attempt is to be made to rationalise the method of allocating treatments to batches of raw materials of differing initial quality t to obtain a final quality y. There are three possible treatments 1, 2 and 3 and information on the effectiveness of these treatments has been sought from 30 experimental runs of the process. The results of these runs are shown in table 12.1. If the selling price of a batch of quality y is $g(y)$ and the cost of treatment a is c_a per batch ($a = 1, 2, 3$) what treatment allocation policy should be adopted?

Table 12.1 *Treatments, initial and final qualities in 30 experimental runs of a quality improving process*

Treatment	(Initial quality, final quality)
1	(30.9, 44.2), (35.8, 48.6), (28.2, 44.3), (40.5, 50.0), (23.5, 43.0), (47.4, 52.5), (51.2, 55.0), (43.0, 51.8), (37.7, 49.6), (33.8, 46.1)
2	(33.3, 46.1), (31.3, 46.7), (23.9, 42.7), (42.2, 50.0), (27.4, 45.0), (50.3, 51.0), (35.8, 47.3), (45.7, 51.0), (39.8, 49.1), (37.6, 47.7)
3	(39.8, 46.3), (31.3, 38.6), (41.0, 50.1), (51.2, 57.3), (36.4, 43.1), (45.7, 56.8), (26.0, 37.8), (37.1, 47.3), (43.0, 52.4), (35.2, 45.0)

We shall consider specific forms for $g(y)$ and the c_a later in this chapter. For the moment we are concerned solely with the nature of the treatment allocation problem. In our study of regulation, optimisation, calibration and diagnosis we have already seen that it is necessary to deal with a whole class F of possible future experiments indexed by the elements of T. This feature persists in the present problem since clearly final quality y of a batch can depend appreciably on its initial quality t. There is, however, the additional complication here that final quality may also depend on the treatment a allocated to the batch. Let f_{at} denote the experiment which records the final quality y of a batch of initial quality t subjected to treatment a. Then in analysing this treatment allocation problem we have to envisage the class

$$F = \{f_{at} : a \in A, t \in T\} \tag{12.1}$$

of future experiments, where $A = \{1, 2, 3\}$ is the set of possible treatments and T, the real line say, is the set of possible initial qualities. Then the data of table 12.1 can be expressed in the form

$$z = \{(a_i, t_i, x_i) : i = 1, \ldots, 30\},$$

where the triplet (a, t, x) typically denotes the treatment, initial quality and final quality of a batch. The informative experiment e thus takes the form

$$e = \{f_{a_1 t_1}, \ldots, f_{a_{30} t_{30}}\}.$$

Our treatment allocation problem can then be framed in the following terms. A new batch of initial quality t awaits treatment. On the basis of the information contained in e which of the three future experiments f_{1t}, f_{2t}, f_{3t} is it best to conduct? The term 'best' has clearly to be interpreted here in terms of some kind of utility structure, constructed from the information about selling price and costs. Since the advantage of increasing quality from t to y is $g(y) - g(t)$, and since the cost of treatment a is c_a, we can associate with each triplet (a, t, y) the utility $U(a, t, y)$ of transforming a batch of initial quality t to final quality y by way of treatment a. Then

$$U(a,t,y) = g(y) - g(t) - c_a \quad (a \in A, t \in T, y \in Y). \tag{12.2}$$

Example 12.2

Treatment of a skin allergy. In a clinical trial to compare the effectiveness of two barrier creams in the prevention of recurrence of a certain skin allergy 100 out of 200 previous sufferers were chosen at random and allocated to cream 1, the remainder being assigned to cream 2. Table 12.2 summarises the results of this trial for male and female sufferers separately.

Table 12.2 *Responses of 200 cases of skin allergy to the two barrier creams*

Cream	Response No recurrence	Some recurrence
1	28M, 22F	21M, 29F
2	18M, 32F	35M, 15F

M denotes male, F denotes female

In this example there are just two possible treatments, cream 1 and 2, so that we may take $A = \{1, 2\}$. From the information we have the only possible indicator of appropriate treatment is sex, so that $T = \{M, F\}$. The response to treatment is measured simply in terms of success (no recurrence) or failure (some recurrence); we then take $Y = \{0, 1\}$, where 0 denotes a failure, 1 a success. With this specification the future experiment f_{1M}, for instance, records whether or not there is a recurrence of the skin allergy for a male patient using cream 1. The class F consists of four possible experiments $f_{1M}, f_{1F}, f_{2M}, f_{2F}$, and the informative experiment e consists of 200 independent such experiments, 49 of type f_{1M}, 51 of type f_{1F}, 53 of type f_{2M} and 47 of type f_{2F}. Again the data of table 12.2 can be expressed in the form

$$\mathbf{z} = \{(a_i, t_i, x_i) : i = 1, \ldots, 200\}.$$

The treatment allocation problem, say for a new male sufferer, then consists of deciding whether the future experiment to be performed should be f_{1M} or f_{2M}.

12.2 The prognostic distributions and utility structure

Suppose that for the class F of experiments as defined in (12.1) we can postulate some parametric family of distributions indexed by a parameter $\theta \in \Theta$. More precisely, for the future experiment f_{at} the possible density functions on Y are

$$p(y \mid a, t, \theta) \quad (\theta \in \Theta). \tag{12.3}$$

For instance, in example 12.1 we may consider postulating

$$p(y\,|\,a, t, \boldsymbol{\theta}) = \text{No}(\alpha_a + \beta_a t, \tau_a) \quad (a = 1, 2, 3), \tag{12.4}$$

the assumption that for each treatment there is a normal linear regression of final quality y on initial quality t. This has the form (12.3) with $\boldsymbol{\theta} = (\alpha_1, \beta_1, \tau_1, \alpha_2, \beta_2, \tau_2, \alpha_3, \beta_3, \tau_3)$. We can then regard the duty of the informative experiment with its data of the form

$$\mathbf{z} = \{(a_i, t_i, x_i) : i = 1, \ldots, n\} \tag{12.5}$$

to transform some prior assessment $p(\boldsymbol{\theta})$ into a posterior assessment $p(\boldsymbol{\theta}\,|\,\mathbf{z})$. It is then natural to use this assessment to arrive at the predictive distribution for the future experiment f_{at}:

$$p(y\,|\,a, t, \mathbf{z}) = \int_{\Theta} p(y\,|\,a, t, \boldsymbol{\theta})\, p(\boldsymbol{\theta}\,|\,\mathbf{z})\, \mathrm{d}\boldsymbol{\theta}. \tag{12.6}$$

Our interest in (12.6) is to assess the effect of treatment a on an individual unit in present or initial state t, and in order to choose between treatments we need to obtain such a predictive distribution for each possible treatment. In order to emphasise the dependence of these predictive distributions on treatment we term (12.6) the *prognostic distribution* for treatment a with respect to initial state t.

If there is a well specified utility structure $U(a, t, y)$, assigning a utility to each possible triplet (a, t, y), then the treatment allocation problem can be simply resolved. For an individual unit with indicator t we can evaluate for each possible treatment a the expected utility

$$U(a, t) = \int_{Y} U(a, t, y)\, p(y\,|\,a, t, \mathbf{z})\, \mathrm{d}y, \tag{12.7}$$

expectation being taken with respect to the appropriate prognostic distribution. The optimum treatment $a^*(t)$ corresponding to indicator t maximises this expectation:

$$U\{a^*(t), t\} = \max_{a \in A} U(a, t). \tag{12.8}$$

12.3 Two applications

We are now in a position to apply the decisive theory approach of the preceding section to the treatment allocation problems of § 12.1.

Example 12.1 (continued)

The scatter diagram of fig. 12.1 shows that the model suggested in § 12.2 is not an unreasonable one, so that we set

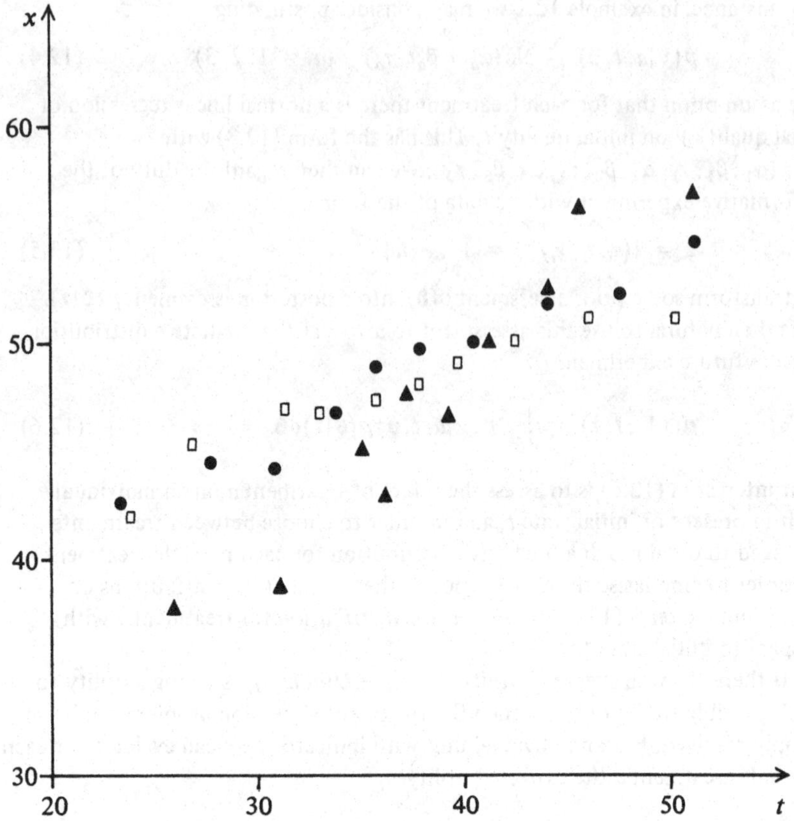

Fig. 12.1 Scatter diagram for quality improving data.
● Treatment 1
□ Treatment 2
▲ Treatment 3

$$p(y\,|\,a, t, \boldsymbol{\theta}) = \text{No}(\alpha_a + \beta_a t, \tau_a).$$

The regression-type computations for each treatment are shown in table 12.3.

Each of the three prognostic distributions is then easily obtained on the basis of vague prior information by a straightforward application of case 5 of table 2.3 along the lines set out in §2.5 exemplified in §10.3. Note that relevant information on $(\alpha_a, \beta_a, \tau_a)$ in the data of table 12.1 comes only from the 10 results on treatment a. The prognostic distributions thus take the form

$$p(y\,|\,a, t, \mathbf{z}) = \text{St}\left[n_a - 2, \hat{\alpha}_a + \hat{\beta}_a t, \left(1 + \frac{1}{n_a} + \frac{(t - \bar{t}_a)^2}{S_a(\mathrm{t}, \mathrm{t})}\right)\frac{v_a}{n_a - 2}\right]$$

with mean $\hat{\alpha}_a + \hat{\beta}_a t$ and are set out in table 12.4.

Table 12.3 *Regression calculations for the three quality-improving treatments*

	Treatment a		
	1	2	3
n_a	10	10	10
$\hat{\alpha}_a$	31.51	35.89	12.52
$\hat{\beta}_a$	0.457	0.321	0.904
$\bar{t}_a$	37.20	36.73	38.67
$S_a(t, t)$	666.72	598.48	466.38
v_a	5.497	3.497	32.633

Table 12.4 *Prognostic distributions for quality improvement*

Treatment	Prognostic distribution $p(y \mid a, t, z) = \mathrm{St}(k, b, c)$		
a	k	b	c
1	8	$31.5 + 0.457\,t$	$0.756 + 0.00103\,(t - 37.20)^2$
2	8	$35.9 + 0.321\,t$	$0.481 + 0.000730\,(t - 36.73)^2$
3	8	$12.5 + 0.904\,t$	$4.487 + 0.00875\,(t - 38.67)^2$

To resolve this treatment allocation problem we require to know the treatment costs c_a $(a = 1, 2, 3)$ and the form of $g(y)$ in (12.2). We examine the problem for two cases.

(i) $c_1 = 4, \quad c_2 = 5, \quad c_3 = 3$ and $g(y) = y$,

so that the selling price is directly proportional to the quality. Then

$$\begin{aligned} U(a, t) &= \int_{-\infty}^{\infty} y p(y \mid a, t, z) \mathrm{d}y - t - c_a \\ &= \hat{\alpha}_a + \hat{\beta}_a t - t - c_a \\ &= \hat{\alpha}_a - c_a - (1 - \hat{\beta}_a)\, t. \end{aligned} \tag{12.9}$$

Fig. 12.2 shows the graphs of $U(a, t)$ plotted against t for each of the three treatments. The optimum treatment allocation rule must then clearly take the form: the optimum treatment is

2 $(t \leqslant 24.8)$,

1 $(24.8 < t \leqslant 40.2)$,

3 $(t > 40.2)$.

(ii) $c_1 = 0.5, \quad c_2 = 0.3, \quad c_3 = 0.6,$ with

$$g(y) = \begin{cases} 2 & (y \geqslant 48), \\ 1 & (y < 48), \end{cases}$$

so that the selling price depends only on whether the quality reaches a standard, 48, or not. Clearly for a batch with $t \geqslant 48$ the standard has already been attained

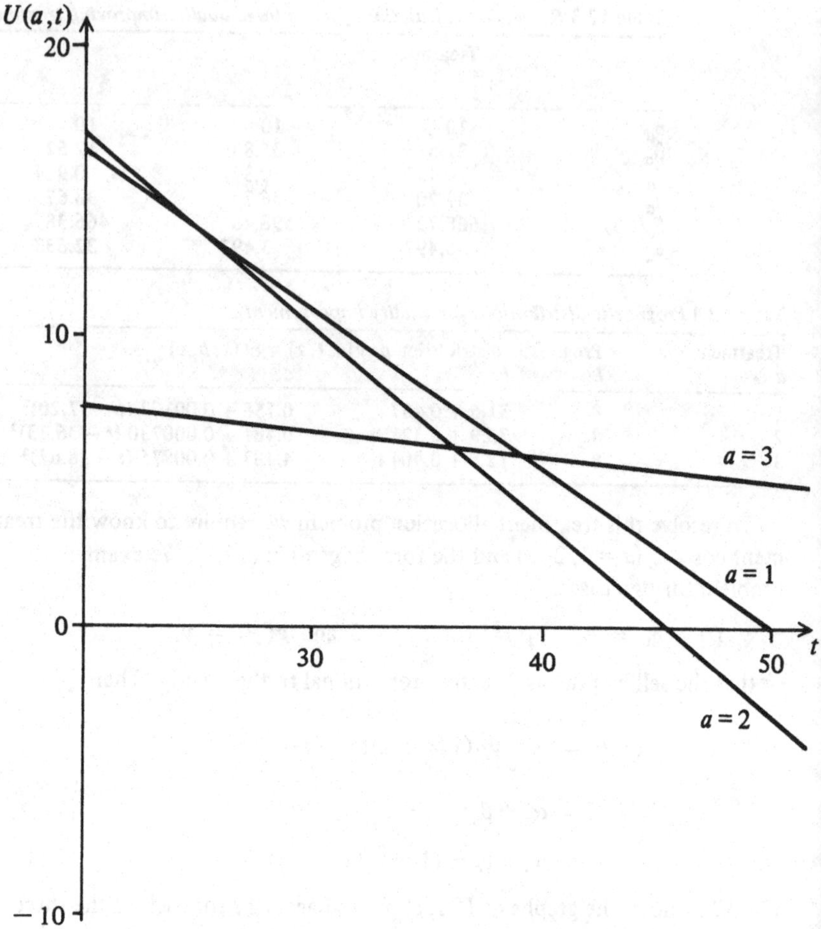

Fig. 12.2 Graphs of $U(a, t)$ plotted against t for each of the three treatments: case (i).

in the raw state and there is thus no sense in treating the batch. We therefore now allow the possibility of a fourth 'treatment' $a = 0$, corresponding to the action of no treatment. Since treatment 0 is clearly optimum for $t \geqslant 48$ we can restrict comparison of the four treatments to the interval $t < 48$. Then for $a = 1, 2, 3,$

$$U(a,t) = 2\int_{48}^{\infty} p(y\,|\,a,t,z)\,dy + \int_{-\infty}^{48} p(y\,|\,a,t,z)\,dy - g(t) - c_a$$

$$= -c_a + \int_{48}^{\infty} p(y\,|\,a,t,z)\,dy \quad (t < 48), \qquad (12.10)$$

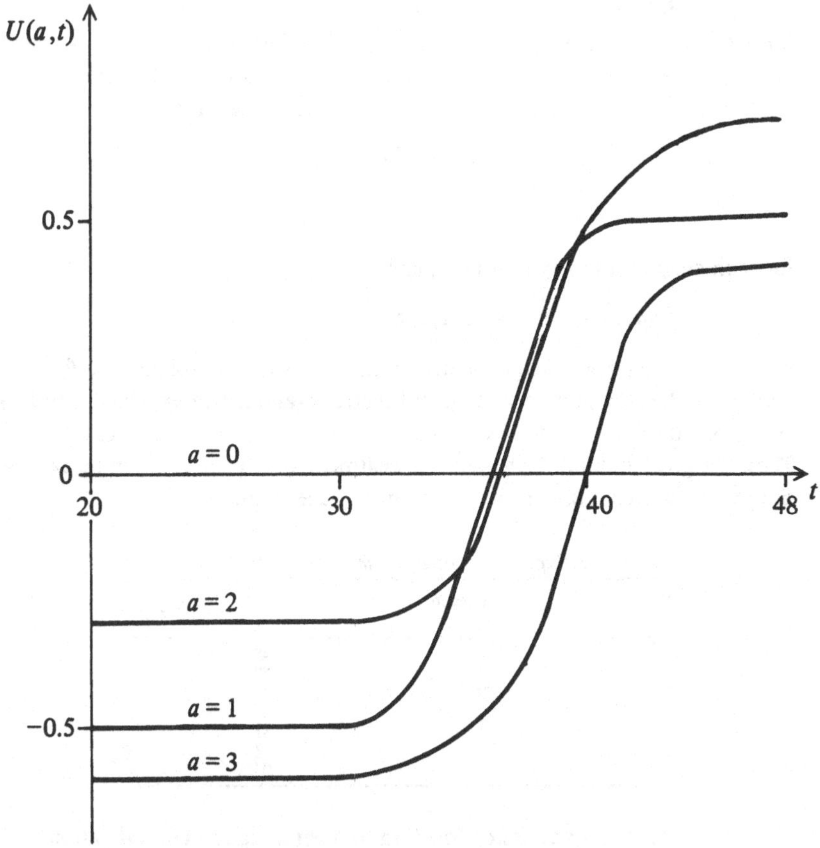

Fig. 12.3 Graphs of $U(a, t)$ plotted against t for each of the four treatments: case (ii).

and

$$U(0, t) = 0 \quad (t < 48). \tag{12.11}$$

The integral in (12.10) can be readily evaluated in terms of the incomplete beta function by (A26) of appendix I (and its simple extension for the case $a < b$). Again the simplest presentation of the solution is in graphical form. In fig. 12.3 the four graphs of $U(a, t)$ against t are shown and from this the following optimum solution emerges: the optimum treatment is

$0 \quad (t \leqslant 36.1)$,
$1 \quad (36.1 < t \leqslant 39.1)$,
$2 \quad (39.1 < t < 48)$,
$0 \quad (48 \leqslant t)$.

Example 12.2 (continued)

If the two creams have the same cost, which can then be set equal to 0, and if the utility structure simply records 1 for a success and 0 for a failure the resolution of the treatment allocation problem is almost trivial. For

$$U(a,t,y) = \begin{cases} 1 & (y = 1), \\ 0 & (y = 0) \end{cases}$$

for each cream a and for each sex t, and so

$$U(a,t) = p(y = 1 \mid a,t,z).$$

If we use vague priors independently on the four success probabilities θ_{at} then, from table 2.3, the predictive density functions (and hence expected utilities) are obtained simply as the observed success rates in the four (a, t) categories. These are given in table 12.5, and the treatment allocation rule can be expressed simply as: for males use cream 1, for females use cream 2.

Table 12.5 *Expected utilities for skin allergy problem*

Sex t	Cream a	$p(y = 1 \mid a, t, z)$
M	1	$\frac{28}{49}$
M	2	$\frac{18}{53}$
F	1	$\frac{22}{51}$
F	2	$\frac{32}{47}$

12.4 Treatment allocation to meet a required specification

One form of problem, considered by Guttman and Tiao (1964) under the heading of 'selecting a best population', is trivially resolved in terms of the treatment allocation theory already developed. Instead of formulating their problem in terms of allocating treatments Guttman and Tiao regard the elements of A as different populations with the associated problem of selecting which of the possible populations is best for a particular utility specification. Moreover in their formulation no initial state t of a population is envisaged but we shall retain this aspect in our presentation here. The main feature in such problems is the simple utility structure which records success and utility 1 for a treated individual unit whose characteristic y or state after treatment falls in a prescribed subset S of Y, and which records failure and utility 0 otherwise. Thus for the case where treatment costs are equal we may set

$$U(a,t,y) = \begin{cases} 1 & (y \in S), \\ 0 & (y \notin S), \end{cases}$$

whatever the initial state t of the unit and whatever the treatment a applied to the unit. For example the units may be electrical components which will operate satisfactorily within a certain amperage range, but fail to function or burn out otherwise.

With this utility structure

$$U(a, t) = \int_S p(y \mid a, t, z)\,dy$$

so that, in terms of the terminology of §4.2, the optimum treatment is that which results in S providing maximum Bayesian cover as computed from the predictive density function. Indeed the analysis here can be regarded as a special case of (12.2) where no treatment is undertaken if $t \in S$ and where

$$g(y) = \begin{cases} 1 & (y \in S), \\ 0 & (y \notin S), \end{cases}$$

and $c_a = 0$ for every a. Thus no new technical problems arise.

History

While treatment allocation problems are an obvious field of application for prediction analysis Guttman and Tiao (1964) appear to have been the first authors to recognise this explicitly. Aitchison (1970) considers a wider range of problems and also investigates the feasibility of attempting to estimate the undeclared utility structure of a treatment allocator from a sequence of treatment allocations made by the allocator.

Problems

12.1 Reconsider example 12.1 with $c_1 = 7, c_2 = 5, c_3 = 2$ and

$$g(y) = 25 - (y - 50)^2.$$

12.2 Reconsider example 12.2 with $c_1 = c_2 = c_3 = 0.2$ and

$$g(y) = \begin{cases} 1 & (44 \leqslant y \leqslant 50), \\ 0 & \text{otherwise}, \end{cases}$$

for the cases where
(i) one of the treatments 1, 2, 3 must be used,
(ii) the possibility of no treatment is allowed.

12.3 A suggested model to explain the variability of the responses of individual units in different initial states and subjected to different treatments is as follows.

For any specified treatment there is a minimum response which is an unknown linear function of initial state and the magnitude of the response in excess of this minimum has an exponential distribution. The table below shows the responses of 75 individual units randomly allocated to the three treatments. The three treatments have equal costs and treatment of a unit is successful if and only if the response is in the interval (12, 25). What treatment allocation policy would you advise?

Indicator	Response to treatment 1	2	3
0	3.3	20.1	12.4
1	5.6	32.0	17.4
2	4.4	19.7	12.8
3	2.9	17.6	12.5
4	4.0	15.5	19.6
5	3.7	24.0	13.8
6	5.7	17.4	17.2
7	12.0	14.5	15.5
8	12.9	19.7	13.0
9	10.3	22.2	16.2
10	11.1	12.5	19.1
11	13.2	15.5	14.8
12	13.0	19.1	26.7
13	14.8	11.0	15.8
14	16.9	13.9	19.5
15	16.9	14.6	16.1
16	21.6	7.6	16.2
17	19.0	7.1	19.0
18	21.1	19.5	17.0
19	23.5	26.2	19.4
20	24.4	8.7	21.1
21	22.1	10.5	19.7
22	28.9	12.6	25.4
23	24.4	4.7	22.6
24	27.6	2.7	21.8
25	31.7	2.8	29.7

APPENDIX I

Notation for standard distributions

The standard distributions are listed in increasing order of dimensionality and in alphabetical order for a given dimensionality. For each distribution the effective sample space or domain of non-zero probability is specified and any restrictions on the indexing parameters are stated.

The following notation is used in the table.

$\mathbf{R}^d$: d-dimensional real space

$\mathcal{S}^d$: the space of positive definite symmetric matrices, defined as the subspace ($\subset \mathbf{R}^{d(d+1)/2}$) of points

$$(w_{11}, w_{12}, \ldots, w_{1d}, w_{22}, \ldots, w_{2d}, w_{33}, \ldots, w_{3d}, \ldots, w_{dd})$$

for which the symmetric matrix $\mathbf{w} = [w_{ij}]$ is positive definite.

$\mathbf{1}$: d-dimensional vector of units.

$\mathbf{I}_d$: d-dimensional identity matrix.

Γ and B denote the gamma and beta functions as usually defined and related by

$$B(g, h) = \frac{\Gamma(g)\Gamma(h)}{\Gamma(g+h)} \quad (g>0, h>0). \tag{A1}$$

Γ_d and B_d are the Siegel (1935) generalisations of the gamma and beta functions defined by

$$\Gamma_d(g) = \pi^{d(d-1)/4}\,\Gamma(g)\Gamma(g-\tfrac{1}{2}) \ldots \Gamma\{g-\tfrac{1}{2}(d-1)\} \quad (g>0), \tag{A2}$$

$$B_d(g, h) = \frac{\Gamma_d(g)\Gamma_d(h)}{\Gamma_d(g+h)} \quad (g>0, h>0). \tag{A3}$$

Two other generalised notations are used, the multinormal coefficient

$$\binom{n}{\mathbf{u}} = \frac{n!}{u_1! \ldots u_d!\,(n-\Sigma u_i)!} \quad (n \text{ positive integer}; u_1, \ldots, u_d \text{ non-negative integers such that } \Sigma u_i \leqslant n) \tag{A4}$$

and the Dirichlet function

$$D(\mathbf{g}, h) = \frac{\Gamma(g_1) \ldots \Gamma(g_d)\Gamma(h)}{\Gamma(\Sigma g_i + h)} \quad (g_i > 0, i = 1, \ldots, d; h > 0). \tag{A5}$$

Distributions within $\mathbf{R}^1$: random variable u; parameter restrictions $b \in \mathbf{R}^1, c > 0, g > 0, h > 0, k > 0, 0 \leqslant l \leqslant 1, n$ positive integer

Notation	Name	Domain restrictions	Density function
$\mathrm{Be}(g, h)$	Beta	$0 < u < 1$	$\dfrac{u^{g-1}(1-u)^{h-1}}{\mathrm{B}(g, h)}$
$\mathrm{BeBi}(n, g, h)$	Beta–binomial	$u = 0, 1, \ldots, n$	$\dbinom{n}{u}\dfrac{\mathrm{B}(g+u, h+n-u)}{\mathrm{B}(g, h)}$
$\mathrm{Bi}(n, l)$	Binomial	$u = 0, 1, \ldots, n$	$\dbinom{n}{u} l^u (1-l)^{n-u}$
$\mathrm{Ch}(g, h)$	Scaled chi-squared	$u > 0$	$\dfrac{(\frac{1}{2}h)^{g/2} u^{(g/2)-1} \exp(-\frac{1}{2}hu)}{\Gamma(\frac{1}{2}g)}$
$\mathrm{El}(b, c)$	Exponential left-sided	$u < b$	$c \exp\{-c(b-u)\}$
$\mathrm{Er}(b, c)$	Exponential right-sided	$u > b$	$c \exp\{-c(u-b)\}$
$\mathrm{Ex}(h)$	Exponential	$u > 0$	he^{-hu}
$\mathrm{Ga}(g, h)$	Gamma	$u > 0$	$\dfrac{h^g u^{g-1} \exp(-hu)}{\Gamma(g)}$

Notation	Name	Domain restrictions	Density function
Ge(l)	Geometric	$u = 0, 1, 2, \ldots$	$l^{u-1}(1-l)$
InBe(k, g, h)	Inverse–beta	$u > 0$	$\dfrac{h^g u^{k-1}}{B(k,g)(h+u)^{k+g}}$
NeBi(n, l)	Negative–binomial	$u = 0, 1, 2, \ldots$	$\binom{n+u-1}{u} l^u (1-l)^n$
No(b, c)	Normal	$u \in \mathbf{R}^1$	$(2\pi)^{-1/2} c^{1/2} \exp\{-\tfrac{1}{2}c(u-b)^2\}$
Pa(g, h)	Pareto	$u > g$	$\dfrac{hg^h}{u^{h+1}}$
Po(h)	Poisson	$u = 0, 1, 2, \ldots$	$\dfrac{\exp(-h)h^u}{u!}$
Si(k, g, h)	Siegel	$u > 0$	$\dfrac{u^{(g/2)-1}}{B(\tfrac{1}{2}k, \tfrac{1}{2}g)\, h^{g/2}(1+h^{-1}u)^{(k+g)/2}}$
St(k, b, c)	Student	$u \in \mathbf{R}^1$	$\dfrac{1}{B(\tfrac{1}{2}, \tfrac{1}{2}k)(kc)^{1/2}\{1+(kc)^{-1}(u-b)^2\}^{(k+1)/2}}$

Inter-relationships of distributions in $\mathbf{R}^1$

$$\mathrm{Ch}(g,h) = \mathrm{Ga}(\tfrac{1}{2}g, \tfrac{1}{2}h) \qquad \text{(A6)}$$

$$\mathrm{Er}(0,c) = \mathrm{Ex}(c) \qquad \text{(A7)}$$

$$\mathrm{Ex}(h) = \mathrm{Ga}(1,h) \qquad \text{(A8)}$$

$$\mathrm{Ge}(l) = \mathrm{NeBi}(1,l) \qquad \text{(A9)}$$

$$\mathrm{Si}(k,g,h) = \mathrm{InBe}(\tfrac{1}{2}g, \tfrac{1}{2}k, h) \qquad \text{(A10)}$$

Relationships of distributions in $\mathbf{R}^1$ *to* N, χ^2, *t and* F *notation*

$\mathrm{No}(b,c) = \mathrm{N}(b, 1/c)$, a normal distribution with mean b and variance $1/c$. (A11)

$\mathrm{Ch}(g,1) = \chi^2(g)$, a chi-squared distribution with g degrees of freedom. (A12)

$\mathrm{St}(k,0,1) = \mathrm{t}(k)$ a t-distribution with k degrees of freedom. (A13)

$\mathrm{Si}(k,g,k/g) = \mathrm{F}(k,g)$, an F-distribution with k and g degrees of freedom. (A14)

Quantiles

The q-quantile of any one-dimensional distribution is defined as

(i) the value $*$ such that $\int_{u<*} p(u)\mathrm{d}u = q$ for continuous distributions;

(ii) the value $*$ such that $\begin{cases} \sum_{u<*} p(u) \leqslant q, \\ \sum_{u\leqslant *} p(u) > q, \end{cases}$ for discrete distributions.

We shall denote the q-quantiles for particular distributions in the following simple way:

$\mathrm{BeBi}(n,g,h;q)$, $\mathrm{Po}(h;q)$, $\mathrm{No}(b,c;q)$, etc.

In particular as special cases

$$\mathrm{Ch}(g,1;q) = \chi^2(g;q), \qquad \text{(A15)}$$

$\mathrm{No}(0,1;q) = \Phi^{-1}(q)$, where Φ is the N(0, 1) distribution function, (A16)

$$\mathrm{St}(k,0,1;q) = \mathrm{t}(k;q), \qquad \text{(A17)}$$

$$\mathrm{Si}(k,g,k/g;q) = \mathrm{F}(k,g;q). \qquad \text{(A18)}$$

Cumulative Distributions

Incomplete beta function – as tabulated by Pearson (1934)

$$I_a(g,h) = \int_0^a \frac{u^{g-1}(1-u)^{h-1}}{\mathrm{B}(g,h)}\,du. \tag{A19}$$

Incomplete gamma distribution – as tabulated by Pearson (1922)

$$J_a(g) = \int_0^a \frac{u^{g-1}\exp(-u)}{\Gamma(g)}\,du. \tag{A20}$$

Hypergeometric cumulative distribution – as tabulated by Lieberman and Owen (1961)

$$P_{\mathrm{hy}}(N,n,k,a) = \sum_{u=\max(0,n+k-N)}^{a} \frac{k!\,n!}{(k-u)!\,(n-u)!\,u!} \times \frac{(N-k)!\,(N-n)!}{N!\,(N-k-n+u)!}$$

where $N, n, k = 0, 1, 2, \ldots; a \leqslant n \leqslant N, a \leqslant k \leqslant N.$ (A21)

Some useful relationships

$$\sum_a^n \mathrm{Bi}(n,l) = I_l(a, n-a+1) \qquad (a>0) \quad \text{(A22)}$$

$$\sum_a^\infty \mathrm{NeBi}(n,l) = I_l(a,n) \qquad (a>0) \quad \text{(A23)}$$

$$\int_a^\infty \mathrm{InBe}(k,g,h) = I_{h/(h+a)}(g,k) \qquad (a>0) \quad \text{(A24)}$$

$$\int_a^\infty \mathrm{Si}(k,g,h) = I_{h/(h+a)}(\tfrac{1}{2}k, \tfrac{1}{2}g) \qquad (a>0) \quad \text{(A25)}$$

$$\int_a^\infty \mathrm{St}(k,b,c) = \tfrac{1}{2}I_{[1+(kc)^{-1}(a-b)^2]^{-1}}(\tfrac{1}{2}k, \tfrac{1}{2}) \qquad (a>b) \quad \text{(A26)}$$

$$\sum_a^\infty \mathrm{Po}(h) = J_h(a) \qquad (a>0) \quad \text{(A27)}$$

$$\int_0^a \mathrm{Ch}(g,h) = J_{\frac{1}{2}ha}(\tfrac{1}{2}g) \qquad (a>0) \quad \text{(A28)}$$

$$\int_0^a \mathrm{Ga}(g,h) = J_{ha}(g) \qquad (a>0) \quad \text{(A29)}$$

$$\sum_0^a \mathrm{BeBi}(n,g,h) = P_{\mathrm{hy}}(n+g+h-1, a+g, n, a) \qquad (a>0) \quad \text{(A30)}$$

Distributions within $\mathbf{R}^2$: random vector (u, w); parameter restrictions $b \in \mathbf{R}^1, c > 0, g > 0, h > 0, k > 0$.

Notation	Name	Domain restrictions	Density function
$\text{ElGa}(b, c, g, h)$	Exponential (left-sided)-gamma	$u < b$	$p(u \mid w)$ is $\text{El}(b, cw)$
		$w > 0$	$p(w)$ is $\text{Ga}(g, h)$
$\text{ErGa}(b, c, g, h)$	Exponential (right-sided)-gamma	$u > b$	$p(u \mid w)$ is $\text{Er}(b, cw)$
		$w > 0$	$p(w)$ is $\text{Ga}(g, h)$
$\text{NoCh}(b, c, g, h)$	Normal-scaled chi-squared	$u \in \mathbf{R}^1$	$p(u \mid w)$ is $\text{No}(b, cw)$
		$w > 0$	$p(w)$ is $\text{Ch}(g, h)$
$\text{StSi}(k; b, c; g, h)$	Student–Siegel	$u \in \mathbf{R}^1$, $w > 0$	$\dfrac{w^{(g/2)-1}}{D(\frac{1}{2}, \frac{1}{2}k, \frac{1}{2}g)(kc)^{1/2} h^{g/2} \{1 + (kc)^{-1}(u - b)^2 + h^{-1}w\}^{(k+g+1)/2}}$

Distributions within $\mathbf{R}^d$: random vector $\mathbf{u}$; parameter restrictions $\mathbf{b} \in \mathbf{R}^d$, $\mathbf{c} \in \mathcal{S}^d$, $g_i > 0$ $(i = 1, 2, \dots, d)$, $h > 0$, $k > d - 1$, $0 \leqslant l_i \leqslant 1$ $(i = 1, 2, \dots, d)$, $\Sigma l_i \leqslant 1$, n positive integer

Notation	Name	Domain restrictions	Density function
$\text{Di}(\mathbf{g}, h)$	Dirichlet	$0 \leqslant u_i \leqslant 1$ $(i = 1, 2, \dots, d)$ $\Sigma u_i \leqslant 1$	$\dfrac{u_1^{g_1-1} \dots u_d^{g_d-1}(1 - \Sigma u_i)^{h-1}}{D(\mathbf{g}, h)}$
$\text{Di}^*(\mathbf{g}, h)$	Ordered Dirichlet	$0 \leqslant u_1 \leqslant u_2 \leqslant \dots \leqslant u_d \leqslant 1$	$\dfrac{u_1^{g_1-1}(u_2 - u_1)^{g_2-1} \dots (u_d - u_{d-1})^{g_d-1}(1 - u_d)^{h-1}}{D(\mathbf{g}, h)}$
$\text{DiMu}(n, \mathbf{g}, h)$	Dirichlet–multinomial	$u_i = 0, 1, \dots, n$ $(i = 1, 2, \dots, d)$ $\Sigma u_i \leqslant n$	$\dbinom{n}{\mathbf{u}} \dfrac{D(\mathbf{g} + \mathbf{u}, h + n - \Sigma u_i)}{D(\mathbf{g}, h)}$
$\text{Mu}(n, \mathbf{l})$	Multinomial	$u_i = 0, 1, \dots, n$ $(i = 1, 2, \dots, d)$ $\Sigma u_i \leqslant n$	$\dbinom{n}{\mathbf{u}} l_1^{u_1} \dots l_d^{u_d}(1 - \Sigma l_i)^{n - \Sigma u_i}$
$\text{NeMu}(n, \mathbf{l})$	Negative–multinomial	$u_i = 0, 1, 2, \dots$ $(i = 1, 2, \dots, d)$	$\dbinom{n + \Sigma u_i - 1}{\mathbf{u}} l_1^{u_1} \dots l_d^{u_d}(1 - \Sigma l_i)^n$
$\text{No}_d(\mathbf{b}, \mathbf{c})$	Normal	$\mathbf{u} \in \mathbf{R}^d$	$(2\pi)^{-d/2}\lvert\mathbf{c}\rvert^{1/2} \exp\{-\tfrac{1}{2}(\mathbf{u} - \mathbf{b})'\mathbf{c}(\mathbf{u} - \mathbf{b})\}$
$\text{St}_d(k, \mathbf{b}, \mathbf{c})$	Student	$\mathbf{u} \in \mathbf{R}^d$	$\dfrac{1}{D\{\tfrac{1}{2}\mathbf{1}, \tfrac{1}{2}(k - d + 1)\}\lvert k\mathbf{c}\rvert^{1/2}\{1 + (\mathbf{u} - \mathbf{b})'(k\mathbf{c})^{-1}(\mathbf{u} - \mathbf{b})\}^{(k+1)/2}}$

Relationships of distributions within $\mathbf{R}^d$ *and* $\mathbf{R}^1$

When $d = 1$,

$$\mathrm{Di}(g,h) = \mathrm{Be}(g,h) \tag{A31}$$

$$\mathrm{DiMu}(n,g,h) = \mathrm{BeBi}(n,g,h) \tag{A32}$$

$$\mathrm{Mu}(n,l) = \mathrm{Bi}(n,l) \tag{A33}$$

$$\mathrm{NeMu}(n,l) = \mathrm{NeBi}(n,l) \tag{A34}$$

$$\mathrm{No}_d(b,c) = \mathrm{No}(b,c) \tag{A35}$$

$$\mathrm{St}_d(k,b,c) = \mathrm{St}(k,b,c) \tag{A36}$$

Relationships of distributions within $\mathcal{S}^d$ *and* $\mathbf{R}^1$

When $d = 1$,

$$\mathrm{Si}_d\,(k,g,h) = \mathrm{Si}(k,g,h), \tag{A37}$$

$$\mathrm{Wi}_d(g,h) = \mathrm{Ch}(g,h). \tag{A38}$$

Relationships of distributions within $\mathbf{R}^d \times \mathcal{S}^d$ *and* $\mathbf{R}^2$

When $d = 1$,

$$\mathrm{NoWi}_d(b,c,g,h) = \mathrm{NoCh}(b,c,g,h) \tag{A39}$$

$$\mathrm{StSi}_d(k;b,c;g,h) = \mathrm{StSi}(k;b,c;g,h). \tag{A40}$$

Distributions within $\mathcal{S}^d$: random positive definite symmetric matrix $\mathbf{w}$; parameter restrictions $k > d-1, g > d-1, \mathbf{h} \in \mathcal{S}^d$

Notation	Name	Domain restrictions	Density function
$\mathrm{Si}_d(k, g, \mathbf{h})$	Siegel	$\mathbf{w} \in \mathcal{S}^d$	$\dfrac{\lvert\mathbf{w}\rvert^{(g-d-1)/2}}{\mathrm{B}_d(\frac{1}{2}k, \frac{1}{2}g)\lvert\mathbf{h}\rvert^{g/2}\lvert\mathbf{I}_d + \mathbf{h}^{-1}\mathbf{w}\rvert^{(k+g)/2}}$
$\mathrm{Wi}_d(g, \mathbf{h})$	Wishart	$\mathbf{w} \in \mathcal{S}^d$	$\dfrac{\lvert\frac{1}{2}\mathbf{h}\rvert^{g/2}\lvert\mathbf{w}\rvert^{(g-d-1)/2}\exp\{-\frac{1}{2}\mathrm{tr}\,\mathbf{hw}\}}{\Gamma_d(\frac{1}{2}g)}$

Distributions within $\mathbf{R}^d \times \mathcal{S}^d$: random $(\mathbf{u}, \mathbf{w})$; parameter restrictions: $\mathbf{b} \in \mathbf{R}^d, \mathbf{h} \in \mathcal{S}^d, c > 0, g > d-1, \mathbf{c} \in \mathcal{S}^d, k > d-1$

Notation	Name	Domain restrictions	Density function
$\mathrm{NoWi}_d(\mathbf{b}, c, g, \mathbf{h})$	Normal–Wishart	$\mathbf{u} \in \mathbf{R}^d$ $\mathbf{w} \in \mathcal{S}^d$	$p(\mathbf{u}\mid\mathbf{w})$ is $\mathrm{No}_d(\mathbf{b}, c\mathbf{w})$ $p(\mathbf{w})$ is $\mathrm{Wi}_d(g, \mathbf{h})$
$\mathrm{StSi}_d(k; \mathbf{b}, \mathbf{c}; g, \mathbf{h})$	Student–Siegel	$\mathbf{u} \in \mathbf{R}^d$ $\mathbf{w} \in \mathcal{S}^d$	$\dfrac{\Gamma_d(\frac{1}{2}(k+g+1))\lvert\mathbf{w}\rvert^{(g-d-1)/2}}{\Gamma_d(\frac{1}{2}k)\Gamma_d(\frac{1}{2}g)\pi^{d/2}\lvert k\mathbf{c}\rvert^{1/2}\lvert\mathbf{h}\rvert^{g/2}\lvert\mathbf{I}_d + (k\mathbf{c})^{-1}(\mathbf{u}-\mathbf{b})(\mathbf{u}-\mathbf{b})' + \mathbf{h}^{-1}\mathbf{w}\rvert^{(k+g+1)/2}}$

APPENDIX II

Classification of prediction problems

Type of problem	Future experiment	Predictive density function	Action set A	Domain of utility function
Decisive prediction	Single, f	$p(y \mid x)$	Y (point) $\mathcal{Y}$ (set)	$A \times Y$
Simple sampling inspection	Single, f	$p(y \mid x)$	Sampling actions	$A \times Y$
Regulation Optimisation	Class F indexed by T	$p(y \mid t, z)$	T	$A \times Y \times y_0$ $A \times Y$
Calibration Diagnosis	Class F indexed by T — discrete or continuous / discrete	$p(u \mid y, z)$ (calibrative) $p(u \mid y, z)$ (diagnostic)	T	$A \times T$
Treatment allocation	Class indexed by initial state T and treatment A	$p(y \mid a, t, z)$ (prognostic)	Treatments	$A \times T \times Y$

Bibliography

In addition to books and research papers referred to directly in the text this Bibliography includes a selection of other references (indicated by an asterisk) to provide the interested reader with material for further study.

Aitchison, J. (1963). *The estimation of design values.* Part I. Research Report, Hospital Engineering Research Unit, Glasgow.

Aitchison, J. (1964). Bayesian tolerance regions. *J.R. Statist. Soc.* B **26**, 161–75.

Aitchison, J. (1966). Expected-cover and linear-utility tolerance intervals. *J.R. Statist. Soc.* B **28**, 57–62.

Aitchison, J. (1970). Statistical problems of treatment allocation. *J.R. Statist. Soc.* A **133**, 206–39.

Aitchison, J. and Brown, J.A.C. (1957). *The Lognormal Distribution,* Cambridge University Press.

Aitchison, J. and Kay, J.W. (1975). Principles, practice and performance in decision making in clinical medicine. *Proceedings of the 1973 NATO Conference on the Role and Effectiveness of Decision Theory.*

Aitchison, J., Habbema, D. and Kay, J.W. (1975). Estimative and predictive diagnosis: a critical comparison. In preparation.

Aitchison, J. and Sculthorpe, Diane (1964). *Design value tables.* Research Report, Hospital Engineering Research Unit, Glasgow.

Aitchison, J. and Sculthorpe, Diane (1965). Some problems of statistical prediction. *Biometrika* **52**, 469–83.

Anderson, T.W. (1958). *An Introduction to Multivariate Statistical Analysis.* New York: Wiley.

Ando, A. and Kaufman, G.M. (1965). Bayesian analysis of the independent normal process – neither mean nor precision known. *J. Amer. Statist. Soc.* **60**, 347–58.

Aoki, M. (1967). *Optimisation of Stochastic Systems.* New York: Academic Press.

Bain, L.J. and Weeks, D.L. (1965). Tolerance limits for the generalized gamma distribution. *J. Amer. Statist. Ass.* **60**, 1142–52.

*Barlow, R.E. and Proschan, F. (1966). Tolerance and confidence limits for classes of distributions based on failure rate. *Ann. Math. Statist.* **37**, 1593–601.

Berkson, J. (1969). Estimation of a linear function for a calibration line: consideration of a recent proposal. *Technometrics* **11**, 649–60.

*Birnbaum, Z.W. and Zuckerman, H.S. (1949). A graphical determination of sample size for Wilks' tolerance limits. *Ann. Math. Statist.* **20**, 313–16.

*Bohrer, R. (1967). A note on tolerance limits with Type I censoring. *Technometrics* **10**, 392.

*Bowden, D.C. (1968). Tolerance interval in regression. *Technometrics* **10**, 207–9.

*Bowker, A.H. (1946). Computation of factors for tolerance limits on a normal distribution when the sample is large. *Ann. Math. Statist.* **17**, 238–40.

*Bowker, A.H. (1947). Tolerance limits for normal distribution. Chapter 2 of *Techniques of Statistical Analysis,* ed. C. Eisenhart, M.W. Hastay and W.A. Wallis, pp. 95–110. Statistical Research Group, Columbia University. New York: McGraw-Hill.

Bowker, A.H. and Lieberman, G.J. (1959). *Engineering Statistics.* Englewood Cliffs: Prentice-Hall.
Bratcher, T.L., Schucany, W.R. and Hunt, H.H. (1971). Bayesian prediction and population size assumptions. *Technometrics* **13**, 678–81.
Brownlee, K.A. (1960). *Statistical Theory and Methodology in Science and Engineering.* New York: Wiley.
*Chernoff, H. and Ray, S.N. (1965). A Bayes sequential sampling inspection plan. *Ann. Math. Statist.* **36**, 1387–407.
Chew, V. (1966). Confidence, prediction and tolerance regions for the multivariate normal distribution. *J. Amer. Statist. Ass.* **61**, 605–17.
*Chew, V. (1968). Simultaneous prediction intervals. *Technometrics* **10**, 323–30.
*Cox, C.P. (1971). Interval estimation for *X*-predictors from linear *Y*-on-*X* regression lines through the origin. *J. Amer. Statist. Ass.* **66**, 749–51.
*Danziger, L. and Davis, S.A. (1964). Tables of distribution-free tolerance limits. *Ann. Math. Statist.* **35**, 1361–5.
Davies, O.L. (1960). *The Design and Analysis of Industrial Experiments.* 2nd edn., Edinburgh: Oliver and Boyd.
De Finetti, B. (1964). Foresight: its logical laws, its subjective sources. In *Studies in Subjective Probability*, ed. H.E. Kyburg and H.E. Smokler, pp. 93–138 New York: Wiley.
De Groot, M.H. (1970). *Optimal Statistical Decisions.* New York: McGraw-Hill.
*De Oliveira, J.T. (1966). Quasi-linearly invariant prediction. *Ann. Math. Statist.* **37**, 1684–7.
Draper, N.R. and Guttman, I. (1968*a*). Some Bayesian stratified two-phase sampling results. *Biometrika* **55**, 131–9.
Draper, N.R. and Guttman, I. (1968*b*). Bayesian stratified two-phase sampling results: *k* characteristics. *Biometrika* **55**, 587–9.
Dunsmore, I.R. (1966). A Bayesian approach to classification. *J.R. Statist. Soc.* B **28**, 568–77.
Dunsmore, I.R. (1968). A Bayesian approach to calibration. *J.R. Statist. Soc.* B **30**, 396–405.
Dunsmore, I.R. (1969). Regulation and optimization. *J.R. Statist. Soc.* B **31**, 160–70.
*Dunsmore, I.R. (1974). The Bayesian predictive distribution in life testing models. *Technometrics* **16**, 455–60.
*Easterling, R.G. and Weeks, D.L. (1970). An accuracy criterion for Bayesian tolerance intervals. *J.R. Statist. Soc.* B **32**, 236–40.
Eisenhart, C. (1939). The interpretation of certain regression methods and their use in biological and industrial research. *Ann. Math. Statist.* **10**, 162–86.
*Eisenhart, C., Hastay, M.W. and Wallis, W.A. (1947). *Techniques of Statistical Analysis.* New York: McGraw-Hill.
Ellison, B.E. (1964). On two-sided tolerance intervals for a normal distribution. *Ann. Math. Statist.* **35**, 762–72.
*Enis, P. and Geisser, S. (1971). Estimation of the probability that $Y < X$. *J. Amer. Statist. Ass.* **66**, 162–8.
*Evans, D.H. (1967). An application of numerical integration techniques to statistical tolerancing. *Technometrics* **9**, 441–56.
*Faulkenberry, G.D. (1970). A note on tolerance limits for a binomial distribution. *Technometrics* **12**, 920–2.
*Faulkenberry, G.D. (1973). A method of obtaining prediction intervals. *J. Amer. Statist. Assoc.* **68**, 433–5.
*Faulkenberry, G.D. and Daly, J.C. (1970). Sample size for tolerance limits on a normal distribution. *Technometrics* **12**, 813–21.
*Faulkenberry, G.D. and Weeks, D.L. (1968). Sample size determination for tolerance limits. *Technometrics* **10**, 343–8.
Ferguson, T.S. (1967). *Mathematical Statistics: A Decision Theoretic Approach.* New York: Academic Press.
Ferriss, J.B., Brown, J.J., Fraser, R., Kay, A.W., Neville, A.M., O'Muircheartaigh, I.G.,

Robertson, J.I.S., Symington, T. and Lever, A.F. (1970). Hypertension with aldosterone excess and low plasma-renin: preoperative distinction between patients with and without adrenocortical tumour. *The Lancet* (1970) **2**, 995–1000.
Fisher, R.A. (1935). The fiducial argument in statistical inference. *Ann. Eugen.* **6**, 391–8.
Fisher, R.A. (1936). The use of multiple measurements in taxonomic problems. *Ann. Eugen.* **7**, 179–88.
*Fraser, D.A.S. (1953*a*). Non-parametric tolerance regions. *Ann. Math. Statist.* **24**, 44–55.
*Fraser, D.A.S. (1953*b*). Characterization of tolerance regions. *Ann. Math. Statist.* **24**, 693 (abstract).
*Fraser, D.A.S. and Guttman, I. (1956). Tolerance regions. *Ann. Math. Statist.* **27**, 162–79.
*Frawley, W.H., Kapadia, C.H., Rao, J.N.K. and Owen, D.B. (1971). Tolerance limits based on range and mean range. *Technometrics* **13**, 651–6.
*Gardiner, D.A. and Hull, N.C. (1966). An approximation to two-sided tolerance limits for normal populations. *Technometrics* **8**, 115–22.
*Gaylor, D.W. and Sweeny, H.C. (1965). Design for optimal prediction in simple linear regression. *J. Amer. Statist. Ass.* **60**, 205–16.
Geisser, S. (1964). Posterior odds for multivariate normal classifications. *J.R. Statist. Soc.* B **26**, 69–76.
Geisser, S. (1971). The inferential use of predictive distributions. In *Foundations of Statistical Inference*, ed. V.P. Godambe and D.A. Sprott, pp. 456–69. Toronto: Holt, Rinehart and Winston.
*Geisser, S. (1974). A predictive approach to the random effect model. *Biometrika* **61**, 101–7.
*Geisser, S. and Desu, M.M. (1968). Predictive zero-mean uniform discrimination. *Biometrika* **55**, 519–24.
*Goldberger, A.S. (1962). Best linear unbiased prediction in the generalized linear regression model. *J. Amer. Statist. Ass.* **57**, 369–75.
*Goodman, L.A. (1953). Parameter-free and non-parametric tolerance limits: the exponential case. *Ann. Math. Statist.* **24**, 139–40 (Abstract).
*Goodman, L.A. and Madansky, A. (1962). Parameter-free and non-parametric tolerance limits: the exponential case. *Technometrics* **4**, 75–95.
Guthrie, D. and Johns, M.V. (1959). Bayes acceptance sampling procedures for large lots. *Ann. Math. Statist.* **30**, 896–925.
*Guttman, I. (1967). The use of the concept of a future observation in goodness-of-fit problems. *J. R. Statist. Soc.* B **29**, 83–100.
Guttman, I. (1970). *Statistical Tolerance Regions: Classical and Bayesian.* London: Griffin.
Guttman, I. and Tiao, G.C. (1964). A Bayesian approach to some best population problems. *Ann. Math. Statist.* **35**, 825–35.
*Hahn, G.J. (1969). Factors for calculating two-sided prediction intervals for samples from a normal distribution. *J. Amer. Statist. Ass.* **64**, 878–88.
*Hahn, G.J. (1970). Additional factors for calculating prediction intervals for samples from a normal distribution. *J. Amer. Statist. Ass.* **65**, 1668–76.
*Hahn, G.J. (1972). Simultaneous prediction intervals for a regression model. *Technometrics* **14**, 203–14.
Halperin, M. (1970). On inverse estimation in linear regression. *Technometrics* **12**, 727–36.
*Hanson, D.L. and Koopmans, L.H. (1964). Tolerance limits for the class of distributions with increasing hazard rates. *Ann. Math. Statist.* **35**, 1561–70.
*Hanson, D.L. and Owen, D.B. (1963). Distribution-free tolerance limits: elimination of the requirement that the cumulative distribution functions be continuous. *Technometrics* **5**, 518–22.
*Hewett, J.E. (1968). A note on prediction intervals based on partial observations in certain life test experiments. *Technometrics* **10**, 850–3.
*Hickman, J.C. (1963). Preliminary regional forecasts of the outcome of an estimation problem. *J. Amer. Statist. Ass.* **58**, 1104–12.

Hoadley, A.B. (1970). A Bayesian look at inverse linear regression. *J. Amer. Statist. Ass.* **65**, 356–69.
*Hora, R.B. and Buehler, R.J. (1967). Fiducial theory and invariant prediction. *Ann. Math. Statist.* **38**, 795–801.
Howe, W.G. (1969). Two-sided tolerance limits for normal populations – some improvements. *J. Amer. Statist. Ass.* **64**, 610–20.
Jeffreys, H. (1961). *Theory of Probability.* 3rd edn., Oxford University Press.
*Jilek, M. and Likar, O. (1959). Coefficients for the determination of one-sided tolerance limits of normal distribution. *Ann. Inst. Statist. Math.* **11**, 45–8.
*Jilek, M. and Likar, O. (1960). Tolerance limits of the normal distribution with known variance and unknown mean. *Austral. J. Statist.* **2**, 78–83.
*John, S. (1963). A tolerance region for multivariate normal distributions. *Sankhya* A **25**, 363–8.
*John, S. (1968). A central tolerance region for the multivariate normal distribution. *J.R. Statist. Soc.* B **30**, 599–601.
Kabe, D.G. (1967). On multivariate prediction intervals for sample mean and covariance based on partial observations. *J. Amer. Statist. Ass.* **62**, 634–7.
*Kalbfleisch, J.D. (1971). Likelihood methods of prediction. In *Foundations of Statistical Inference*, ed. V.P. Godambe and D.A. Sprott, pp. 378–92. Toronto: Holt, Rinehart and Winston.
Kalotay, A.J. (1971). Structural solution to the linear calibration problem. *Technometrics* **13**, 761–9.
*Kemperman, J.H.B. (1956). Generalized tolerance limits. *Ann. Math. Statist.* **27**, 180–6.
Krutchkoff, R.G. (1967). Classical and inverse regression methods in calibration. *Technometrics* **9**, 425–39.
Krutchkoff, R.G. (1968). Letter to the Editor. *Technometrics* **10**, 430–1.
Krutchkoff, R.G. (1969). Classical and inverse regression methods of calibration in extrapolation. *Technometrics* **11**, 605–8.
*Lawless, J.F. (1970). A prediction problem concerning samples from the exponential distribution with application in life testing. *Technometrics* **13**, 725–30.
*Lawless, J.F. (1972). On prediction intervals for samples from the exponential distribution and prediction limits for system survival. *Sankhya* B **34**, 1–14.
*Lawrence, M.J. (1967). Inequalities and tolerance limits for *s*-ordered distributions. *Ann. Math. Statist.* **38**, 1595–6. (abstract).
Lieberman, G.J. (1957). *Table for the determination of two-sided tolerance limits for the normal distribution.* Report no. 373–17 (55), Bureau of Ships, Navy Department, Washington.
*Lieberman, G.J. (1961). Prediction regions for several predictions from a single regression analysis. *Technometrics* **3**, 21–7.
*Lieberman, G.J. and Miller, R.G. (1963). Simultaneous tolerance intervals in regression. *Biometrika* **50**, 155–68.
*Lieberman, G.J., Miller, R.G. and Hamilton, M.A. (1967). Unlimited simultaneous discrimination intervals in regression. *Biometrika* **54**, 133–45.
Lieberman, G.J. and Owen, D.B. (1961). *Tables of the Hypergeometric Probability Distributions.* Stanford: Stanford University Press.
*Lindley, D.V. (1965*a*). *Introduction to Probability and Statistics from a Bayesian Viewpoint, Part I: Probability.* Cambridge University Press.
Lindley, D.V. (1965*b*). *Introduction to Probability and Statistics from a Bayesian Viewpoint, Part II: Inference.* Cambridge University Press.
Lindley, D.V. (1968). The choice of variables in multiple regression. *J. R. Statist. Soc.* B **30**, 31–66.
*Lindley, D.V. (1969). *A Bayesian solution for some educational prediction problems II.* Educational Testing Service Research Bulletin, Princeton.
Lindley, D.V. (1971). *Bayesian Statistics, A Review.* Regional Conference Series in Applied Mathematics. Philadelphia: SIAM.
Lindley, D.V. and Barnett, B.N. (1965). Sequential sampling: two decision problems with

linear losses for binomial and normal random variables. *Biometrika* **52**, 507–32.
*Mack, C. (1969). Tolerance limits for a binomial distribution. *Technometrics* **11**, 201.
Mandel, J. and Linnig, F.J. (1957). Study of accuracy in chemical analysis using linear calibration curves. *Anal. Chem.* **29**, 743–9.
Maritz, J.S. (1970). *Empirical Bayes Model.* London: Methuen.
Martinelle, S. (1970). On the choice of regression in linear calibration: comments on a paper by R.G. Krutchkoff. *Technometrics* **12**, 157–61.
Mises, R. von. (1942). On the correct use of Bayes' formula. *Ann. Math. Statist.* **13**, 156–65.
*Mitra, S.K. (1957). Tables for tolerance limits for a normal population based on sample mean and range or mean range. *J. Amer. Statist. Ass.* **52**, 88–94.
*Murphy, R.B. (1948). Non-parametric tolerance limits. *Ann. Math. Statist.* **19**, 581–9.
Neyman, J. (1962). Two breakthroughs in the theory of statistical decision making. *Rev. Int. Statist. Inst.* **30**, 11–27.
*Noether, G.E. (1951). On a connection between confidence and tolerance intervals. *Ann. Math. Statist.* **22**, 603–4.
Oden, A. (1973). Simultaneous confidence intervals in inverse linear regression. *Biometrika* **60**, 339–43.
*Ott, R.L. and Myers, R.H. (1968). Optical experimental designs for estimating the independent variable in regression. *Technometrics* **10**, 811–23.
Owen, D.B. (1962). *Handbook of Statistical Tables.* Reading, Massachusetts: Addison-Wesley.
*Owen, D.B., Craswell, K.J. and Hanson, D.L. (1964). Non-parametric upper confidence bounds for $Pr(Y < X)$ and confidence limits for $Pr(Y < X)$ when X and Y are normal. *Ann. Math. Statist.* **35**, 457–8 (abstract).
Paulson, E. (1943). A note on tolerance limits. *Ann. Math. Statist.* **14**, 90–3.
Pearson, E.S. and Hartley, H.O. (1966). *Biometrika Tables for Statisticians.* Vol. I (3rd edn.), Cambridge University Press.
Pearson, K. (1922). *Tables of the Incomplete Γ-function.* Cambridge University Press.
Pearson, K. (1934). *Tables of the Incomplete B-function.* Cambridge University Press.
Proschan, F. (1953). Confidence and tolerance intervals for the normal distribution. *J. Amer. Statist. Ass.* **48**, 550–64.
*Quesenberry, C.P. and Gessaman, M.P. (1968). Non-parametric discrimination using tolerance regions. *Ann. Math. Statist.* **39**, 664–73.
Raiffa, H. and Schlaifer, R. (1961). *Applied Statistical Decision Theory.* Harvard University Press.
Resnikoff, G.J. (1962). Tables to facilitate the computation of percentage points of the non-central t-distribution. *Ann. Math. Statist.* **33**, 580–6.
Resnikoff, G.J. and Lieberman, G.J. (1957). *Tables of the Non-central t-distribution.* Stanford: University Press.
*Robbins, H. (1944). On distribution-free tolerance limits in random sampling. *Ann. Math. Statist.* **15**, 214–16.
Robbins, H. (1955). An empirical Bayes approach to statistics. In *Proc. 3rd Berkeley Symposium on Mathematical Statistics and Probability*, ed. J. Neyman, pp. 157–64. Berkeley: University of California Press.
Roberts, H.V. (1965). Probabilistic prediction. *J. Amer. Statist. Ass.* **60**, 50–62.
*Saunders, S.C. (1956). Sequential distribution free tolerance regions. *Ann. Math. Statist.* **27**, 865 (abstract).
*Saunders, S.C. (1960). Sequential tolerance regions. *Ann. Math. Statist.* **31**, 198–216.
Sawagari, Y., Sunahara, Y. and Nakamizo, T. (1967). *Statistical Decision Theory in Adaptive Control Systems.* New York: Academic Press.
Scheffé, H. (1973). A statistical theory of calibration. *Ann. Statist.* **1**, 1–37.
*Scheffé, H. and Tukey, J.W. (1944). A formula for sample sizes for population tolerance limits. *Ann. Math. Statist.* **15**, 217.
*Scott, A.J. and Symons, M.J., (1971). A note on shortest prediction intervals for log-linear regression. *Technometrics* **13**, 889–94.

*Sharpe, K. (1970). Robustness of normal tolerance intervals. *Biometrika* **57**, 71–8.
Shukla, G.K. (1972). On the problem of calibration. *Technometrics* **14**, 547–53.
Siegel, C.L. (1935). Uber die analytische Theorie der quadratischen Formen. *Ann. Maths.* **36**, 527–606. (See also Ando and Kaufman, 1965.)
*Somerville, P.N. (1958). Tables for obtaining non-parametric tolerance limits. *Ann. Math. Statist.* **29**, 599–601.
*Stone, M. (1974). Cross-validatory choice and assessment of statistical predictions. *J. R. Statist. Soc.* B **36**, 111–47.
*Tallis, G.M. (1969). Note on a calibration problem. *Biometrika* **56**, 505–8.
Teicher, H. (1960). On the mixture of distributions. *Ann. Math. Statist.* **31**, 55–73.
Teicher, H. (1961). Identifiability of mixtures. *Ann. Math. Statist.* **32**, 244–8.
Teicher, H. (1963). Identifiability of finite mixtures. *Ann. Math. Statist.* **34**, 1265–9.
Thatcher, A.R. (1964). Relationships between Bayesian and confidence limits for predictions. *J. R. Statist. Soc.* B **26**, 176–92.
*Thoman, D.R., Bain, L.J. and Antle, C.E. (1970). Maximum likelihood estimation, exact confidence limits for reliability and tolerance limits in the Weibull distribution. *Technometrics* **12**, 363–71.
*Tukey, J.W. (1947). Non-parametric estimation, II: statistically equivalent blocks and tolerance regions – the continuous case. *Ann. Math. Statist.* **18**, 529–39.
*Tukey, J.W. (1948). Non-parametric estimation, III: statistically equivalent blocks and multivariate tolerance regions – the discontinuous case. *Ann. Math. Statist.* **19**, 30–9.
*Varde, S.D. (1969). Life testing and reliability estimation for the two-parameter exponential distribution. *J. Amer. Statist. Ass.* **64**, 621–31.
*Wald, A. (1942). Setting of tolerance limits when the sample is large. *Ann. Math. Statist.* **13**, 389–99.
*Wald, A. (1943). An extension of Wilks' method for setting tolerance limits. *Ann. Math. Statist.* **14**, 45–55.
Wald, A. (1950). *Statistical Decision Functions.* New York: Wiley.
Wald, A. and Wolfowitz, J. (1946). Tolerance limits for a normal distribution. *Ann. Math. Statist.* **17**, 208–15.
Wallis, W.A. (1951). Tolerance intervals for linear regression. In *Proc. 2nd Berkeley Symposium on Mathematical Statistics and Probability*, ed. J. Neyman, pp. 43–51. Berkeley: University of California Press.
*Walsh, J.E. (1962). Distribution-free tolerance intervals for continuous symmetrical populations. *Ann. Math. Statist.* **33**, 1167–74.
*Walsh, J.E. (1962). Non-parametric confidence intervals and tolerance regions. Chapter 8 in *Contributions to Order Statistics*, ed. A.E. Sarhan and B.G. Greenbery. New York: Wiley.
*Walsh, J.E. (1962). Some two-sided distribution-free tolerance intervals of a general nature. *J. Amer. Statist. Ass.* **57**, 775–84.
Weissberg, A. and Beatty, G.H. (1960). Tables of tolerance-limit factors for normal distributions. *Technometrics* **2**, 483–500.
Wetherill, G.B. (1966). *Sequential Methods in Statistics.* London: Methuen.
Wetherill, G.B. and Campling, G.E.G. (1966). The decision theory approach to sampling inspection. *J. R. Statist. Soc.* B **28**, 381–416.
Wilks, S.S. (1941). Determination of sample sizes for setting tolerance limits. *Ann. Math. Statist.* **12**, 91–6.
*Wilks, S.S. (1942). Statistical prediction with special reference to the problem of tolerance limits. *Ann. Math. Statist.* **13**, 400–9.
Williams, E.J. (1969*a*). A note on regression methods in calibration. *Technometrics* **11**, 189–92.
Williams, E.J. (1969*b*). Regression methods in calibration problems. *Bull. Int. Statist. Inst.* **43**, 17–28.
*Wilson, A.L. (1967). An approach to simultaneous tolerance intervals in regression. *Ann. Math. Statist.* **38**, 1536–40.
Winkler, R.L. (1972). *An Introduction in Bayesian Inference and Decision.* New York: Holt, Rinehart and Winston.

*Wolfowitz, J. (1946). Confidence limits for the fraction of a normal population which lies between two given limits. *Ann. Math. Statist.* **17**, 483–8.

*Zacks, S. (1970). Uniformly most accurate upper tolerance limits for monotone likelihood ratio families of discrete distributions. *J. Amer. Statist. Ass.* **65**, 307–16.

Zellner, A. and Chetty, V.K. (1965). Prediction and decision problems in regression models from the Bayesian point of view. *J. Amer. Statist. Ass.* **60**, 608–16.

Author index

The full references for the authors listed here are given in the bibliography, which also includes a selection of other references to provide the interested reader with material for further study

Subject index

Example and problem index